U0085093

中英對照
Bilingual Edition by Amanda Chi

蒟蒻

50道DIY蒟蒻料理，讓你吃出漂亮好氣色

纖瘦健康吃

Healthy
High Fiber
Slimming
Low Calories
with Konjac
Easy to Make

齊美玲 著
by Amanda Chi

Eating Konjac = Beautiful

序

漂亮吃蒟蒻

《蒟蒻纖瘦健康吃》是我個人的第二本蒟蒻食譜，距離上一本食譜已經相隔近兩年，這段期間以來，我得到了很多讀者的心情分享，同時也收到了很多的支持與鼓勵，在此先向大家致上衷心的謝意。

以蒟蒻DIY為主題的食譜能夠受到這麼多的迴響，其實是令人感動的，因為，要在忙碌的生活中挪出時間來自己動手做蒟蒻，必定是懷抱著「以愛心為初衷 以照顧家人健康為目標」的心情而來的。我的確在與許多讀者分享玩蒟蒻的樂趣之中學到更多，也因很多人吃蒟蒻改善了身體狀況，而更纖瘦、更健康，獲得許多鼓勵。正因這許多分享與迴響的鼓勵，才促成這本書的出版，我真心希望有更多的人從中領略玩蒟蒻的樂趣並得以改善健康。

為了要回應許多來自國外的讀者要求，本書採中英對照的方式編排，在這裡特別要感謝出版社的費心。近年來蒟蒻的健康價值確實受到了肯定，這也證明了大家愈來愈具有保健的觀念了。

最後建議讀者在開始動手DIY前，一定要先仔細閱讀關於蒟蒻的特性及蒟蒻的基本做法，這樣才能得心應手的成為「蒟蒻玩家」，讓我們一起加油吧！

2004.05

《Healthy Slimming with Konjac》is my second konjac cookbook. It's been two years since the first book was published. During this period, I have received many letters from readers sharing their feelings and giving me much support and encouragement. I would like to offer my sincere gratitude to everybody.

My cookbook on DIY konjac at home made so many echoes. I was really touched by them, because in this busy lifestyle taking your precious time to make DIY konjac must start with the goal of "from a caring heart taking caring of the family." I have learned a lot from my readers about playing with konjac. I am encouraged because many people have improved their health after changing their diet to konjac. These thoughts and encouraging responses stimulated the publishing of this cookbook. I sincerely hope that many more people will improve their health through the fun of making konjac.

In response to all the readers abroad, this cookbook is written in both Chinese and English. Here I would like to thank the publishing company for its effort. Recently the value of konjac for the health has been positively confirmed. This proves that everyone is getting more and more savvy about keeping fit and healthy.

Let me give you some advice before you start to DIY konjac. Read carefully about the characteristics of the konjac and the Basic Preparation of Konjac, so that it will be easier for you to become an excellent konjac player. Let's keep up the good work!

Amanda Chi May 2004

CONTENTS

果凍點心 Jelly Dessert

CONTENTS

簡易小菜 Simple Side Dishes

Konjac

Konjac

瘦身大餐 Diet Meal

WHAT IS KONJAC?
蒟蒻是什麼？

蒟蒻是數年前經由蒟蒻果凍而被多數人認識。但早在數千年前就已經被用來做為食材，在阿嬤的時代被稱為「粉鳥舅」或「日本豆腐」的，就是蒟蒻。一般人對蒟蒻的感覺很陌生，甚至有些人不能接受它特有的氣味及平淡無味，但如果因為這樣就放棄蒟蒻這種高纖低熱量的天然食材，就太可惜了！因為「蒟蒻」無疑是最適合現代人用來照顧全家人的好食物之一，「不好聞的味道」和「不好吃」是可以解決的。烹調蒟蒻料理前，必須先針對蒟蒻的生長環境、製作過程等做一些說明，讓大家對蒟蒻有進一步的了解。

一、所謂「蒟蒻」

蒟蒻學名Amorp Hophallus Konjac C.Koch，天南星科、蛇芋屬、多年生宿根性塊莖草本植物，我們取用的是它地下塊莖的部分。原產地在印度錫蘭等地，而中國大陸的雲南、四川、貴州、湖北等省份，由於溫度、濕度氣候適宜，加上地大、人工成本低等因素，已經成為全世界最大的蒟蒻產地，包括對蒟蒻情有獨鍾的日本人，也在大陸與當地農民「契作」，並且設置廠房製成蒟蒻粉後再回銷日本，致力於研發各式蒟蒻製品行銷全球。

二、製造過程

蒟蒻在種植到三年最適合採收。採收後的蒟蒻塊莖製成蒟蒻粉是經過了去皮→切片→乾燥→磨碎→過篩→空氣分級→蒟蒻粗粉的過程。

從外觀上來看蒟蒻粉的優劣，細度高、無雜質、色澤白、較無腥味是屬於較高等級；而當蒟蒻粉與水接觸時，其分子會因為吸水膨脹而破裂釋出膠體，期間出膠的速度與和水的膨潤倍數也是判定蒟蒻粉品質好壞的標準。一般蒟蒻粉有粗粉、精粉、微粉之分，除了以上所述的差別之外，優質的蒟蒻粉做出來的蒟蒻成品，口感細緻而且透明度、嚼勁也較佳，適合用來烹調。

Konjac became known many years ago through the invention of konjac jelly. In fact, thousands of years ago, it was being used as a food ingredient. In my grandmother's time, konjac was called by various names like Fen Niao Jiu, and "Japanese tofu." Since most people are very unfamiliar with konjac, its particular smell and tasteless flavor are often rejected. However, if you give up this high fiber, low calorie food for those reasons, it will really be a shame! Why? Because konjac is without a doubt one of the best ingredients for taking care of the family stomach. Its odor and blandness problems can be solved easily. Before cooking konjac, its biological background and preparation process must be discussed, so that everyone can have a clearer understanding of it.

I. Background on konjac

The scientific name of konjac is Amorphophallus konjac C.Koch. It is a member of the Araceae family. The root is edible. The plant is native to India, but is grown across southern China, including Szchuan, Yunnan, Kweichou, and Hubei, which have become the largest konjac producing areas due to the warm, moist climate, large tracts of land, and low cost of labor. The Japanese, who are particularly fond of konjac, make contracts with the local farmers in China. They have also built factories to turn the konjac into powder and export it back to Japan, and have devoted themselves to researching konjac products to sell to the whole world.

II. The Process of Making Konjac

After growing for three years the Konjac is harvested. After collecting the konjac root, it is made into konjac powder, through a process involving peeling - slicing - drying - grinding - sifting - and classification, which produce a coarse konjac powder.

Good quality konjac powder has a fine texture, no impurities, a pure white color, and a less unpleasant smell. When konjac powder comes into contact with water, its molecules will expand due to water absorption, releasing the gel. When this is happening, the speed of gelling and the size of expansion are the standards that determine whether the quality of konjac is good or bad. Normally konjac powder is classified into coarse, delicate and superfine varieties. In addition to the differences mentioned above, good quality konjac powder can be made into outstanding konjac products. The transparent texture is exquisite, and given its wonderful chewiness, it is perfect for use in cooking.

SO MANY GOOD THINGS 吃蒟蒻，好處多多 TO SAY ABOUT EATING KONJAC

蒟蒻是符合現代人健康需求的食材之一，常吃蒟蒻優點多多：

1.含有大量的可溶性纖維，幫助排便順暢

蒟蒻的主要成分是水分、纖維、葡甘露聚糖、少量蛋白質、鈣、鐵、磷等，所含的可溶性纖維能幫助腸胃蠕動、促進腸內毒素及細菌的排出。現代人由於生活緊張、壓力大，飲食普遍不夠均衡，所攝取的纖維往往不足，以致罹患便祕的人相當多，而蒟蒻高纖的特性非常適合幫助有這類困擾的人，在晚餐時吃蒟蒻將有助於排便更順暢。

2.熱量低具飽足感

蒟蒻粉在製造過程中已經將大部分的澱粉去除，可以說幾乎不含熱量；同時又因為水分的含量很高，容易讓人有飽足感。現代人常面臨過度飲食造成肥胖的問題，蒟蒻的這項特性，對於想要控制體重的人很有幫助。既可以滿足口腹之慾，卻不會對身體造成過多的負擔，的確是控制

體重的好幫手。

3.葡甘露聚糖可幫助平衡膽酸

　　蒟蒻的成分中最特別的就是葡甘露聚糖，它是一種多醣體，也就是所謂的膳食纖維。除了可在小腸中吸收膽酸、加速代謝有害物質外，加上它不會被人體消化吸收，卻能被腸內細菌分解的特性，同樣也能夠促進腸內廢棄物及有害細菌的排泄。

Konjac is one of the ingredients that completely match the daily health needs of modern people. There are so many good things about Konjac!

1. It contains large amounts of soluble fiber and facilitates smooth bowel movements.

The main ingredients of konjac are liquid, fiber, and glucomannan, along with a little protein, calcium, iron and phosphorus. The fiber is soluble and promotes intestinal movement, releases toxins in the intestines, and releases bacteria. People today, because of their intense, high-pressure lifestyle, consume food and drink that is mostly not balanced enough. Fiber intake is hardly enough, which make many people experience constipation. The fiber in konjac is perfect for helping people who have these kinds of problems. Serving it at dinner will help the bowel move more smoothly.

2. Low calorie, yet filling

During the process of making konjac powder, most of the starch is removed, that it can be said to have no calories at all. At the same time, because liquid content of the liquid is high, it easily makes you feel full. Nowadays we have to face the problem of overeating, which leads to weight problems. The aforementioned characteristics of konjac are of great help to people who are trying to control their weight. It not only satisfies your craving for food, but does not burden the body.

3. Glucomannan helps balance cholic acid

The most unique element of konjac is glucomannan. It is a kind of polysaccharide, also known as dietary fiber. In addition to absorbing cholic acid in the small intestines, it speeds up metabolization of toxins. It will not be absorbed by the digestive system, yet it can be dissolved by bacteria. It also improves the removal of waste and toxic bacteria in intestines.

SMART SELECTION AND KONJAC PRESERVATION
聰明選擇與保存蒟蒻

蒟蒻是健康、優質的好食材，拿來做菜再適合不過了，然而，在選購及保存上，有幾個小訣竅喔！

一.蒟蒻粉的選擇：

如何選擇品質好的蒟蒻粉呢？可用以下的列表來做比較。

二. 蒟蒻的保存：

蒟蒻的保存依蒟蒻粉和蒟蒻製品而有所不同。

1.蒟蒻粉的保存：在不受陽光直射、不受潮的狀況下蒟蒻粉可以保存1～2年。但因蒟蒻粉是天然農產品且不添加任何增粘劑，所以其黏度會隨時間愈久而逐漸降低，因此，當蒟蒻粉愈接近有效期限時，製作時應略減水分，才能維持原來的Q度。

2.蒟蒻製品的保存：依照基本做法已經結成膠體的生蒟蒻，放在密閉度良好、乾淨的保鮮盒裡存放於冰箱的冷藏室中，可保存7~10天。製作完成也就是已經蒸熟去鹼的熟蒟蒻，則必須放入密封盒且泡在冷開水中存放於冰箱冷藏室，可保存7~10天。保存良好的蒟蒻成品應該是表面光滑、有彈性，如果表面出現組織鬆軟、沒彈性、浸泡的水有混濁現象時，表示蒟蒻已經腐壞了。

Konjac is a healthy and excellent ingredient. It is perfect for all sorts of dishes. In purchasing and preserving konjac, here are some things you should pay attention to:

1. Selecting konjac powder: How to select good quality konjac powder? You can make a comparison using the chart below.

2. Storing konjac: Storage of konjac powder and jelly differ as follows:

1. Storing konjac powder: konjac powder can be stored for 1 to 2 years in a dark, dry. Because konjac powder is a natural agricultural product without any additives, its stickiness will decrease as time goes by. So, when the konjac powder is close to the expiration date, decrease the water added, to maintain its original chewiness.

2. Storing konjac gum: Raw konjac jelly made according to the basic methods can be stored in a clean, well-sealed container in the refrigerator for about 7 to 10 days. Steamed konjac with the alkali removed can be stored in a well-sealed container and soaked in cold water in the refrigerator for about 7 to 10 days. Well-preserved konjac jelly should be smooth and shiny on the surface, with a firm texture. If the surface appears soft and loose, the firm texture will disappear and the soaking water will be dirty, meaning that the konjac jelly has gone bad.

	優質頂級蒟蒻微粉 Top Quality Konjac Powder	**一般蒟蒻精粉** Ordinary Konjac Powder
外觀 Appearance	粉末細、顏色白無雜質、較無腥味 a pure white fine powderno impurities or fishy smell	顆粒較粗、顏色微黃含有雜質、腥味較重 Slightly yellow, coarse powderSome impurities and fishy smell
吸水倍數 Level of Expansion After Absorbing Water	20～25倍 expands to 20-25 times original size	10～15倍 expands 15-20 times original size
遇水出膠成型的速度 Speed of Gel Formation After Absorbing Water	6～10 秒鐘 forms gel in 6-10 seconds	8～10 分鐘 forms gel in 5-8 minutes
質地與口感 Texture and Flavor	質地細緻、口感佳 Fine, delicate texture with superior flavor	纖維較粗糙 Coarser fibers

EATING KONJAC EQUALS SMART PLUS HEALTH
聰明搭配健康吃蒟蒻

蒟蒻具高纖、低熱量的營養價值，是最能符合現代人需求的天然食材，單吃蒟蒻難免會厭煩，但是聰明將蒟蒻搭配其他食材一塊烹調，可以吃得更健康又美味！

蒟蒻除了纖維和水分之外，其他營養素如蛋白質、鈣、磷、鉀等的含量並不是很高，因此，如果只是單吃蒟蒻而不同時攝取其他食物，可能會造成營養不足而引起身體不適，這就有違初衷了。所以，可以在三餐當中多多食用蒟蒻，但是一定要與其他具有養分的食物一起搭配。像具有增強免疫力的菇類、蛋白質含量高的豆類、堅果類等，這些都是非常適合和蒟蒻搭配做為烹調的食材。

聰明吃蒟蒻有兩點須特別注意，一．如果對蒟蒻的需求只是著重在低熱量上，在烹調的方式及調味的選擇上就要注意到口味不要過油、過鹹，否則效果可能會打折。二．蒟蒻具有特殊的嚼勁，在食用上儘可能細嚼慢嚥才不會造成腸胃過度的負擔，進而引起不適，而且每次吃蒟蒻的量不要超過100克，蒟蒻吃得太多，同樣也會造成胃不舒服喔！

Konjac is highly nutritious, with high fiber and low calories. It is a natural ingredients that meets the demands of today's consumers. Eating konjac alone might be a bit tiresome, but in fact, konjac can be prepared with other ingredients and for even greater health benefits. What are the best supporting roles for konjac?

Though high in fiber and liquid, protein, calcium, phosphorus and potassium in Konjac are present in only small amounts. Therefore, eating konjac by itself may lead

to nutritional deficiencies. That would be a shame! So, serve more konjac with two meals a day and of course, serve it with other nutritious foods. For example, serve it with ingredients like mushrooms, which can enhance your immunities, and beans, which are high in protein, or nuts. These are all quite perfect with konjac.

There are two points that attention. First, if your need for konjac is based largely on its low calories, then the cooking method and selection of seasonings should have less oil and salt, or the effect will not be as you desire. Secondly, konjac has a unique texture and is very chewy. Chew as gently and slowly as possible when eating, to prevent burdening the stomach and the resultant discomfort. Moreover, keep the serving of konjac below 100 grams per meal. Too much konjac may cause stomach discomfort.

EASY KONJAC DIY
蒟蒻輕鬆DIY

蒟蒻塊自己做

材料及器具：

蒟蒻粉40克、鹼粉4克、水800c.c.、大碗、打蛋器（圖1）

做法：

1.將水倒入大碗中，先放入鹼粉均勻攪拌，待溶解後以打蛋器持續攪動水，使水形成旋渦。

2.將蒟蒻粉倒於漩渦中，快速攪拌6~10秒鐘即停止。將已凝成膠體的蒟蒻靜置約30分鐘使其狀態穩定，即製成生蒟蒻塊。（圖2、3、4）

3.移入電鍋中，外鍋放1杯半水蒸30~40分鐘，取出蒸好的熟蒟蒻浸泡在水中8~12小時，待完全去鹼才能食用，即製成熟蒟蒻塊。（圖5、6）

Konjac Chunks Do It Yourself

INGREDIENTS AND TOOLS

40g konjac powder, 4g alkali powder, 800cc water, large mixing bowl, eggbeater (fig.1)

METHODS

1. With water in the mixing bowl, add alkali and stir until evenly-mixed and dissolved completely. Beat with an eggbeater continuously to form whirlpools.
2. Pour in all the konjac powder. Stir rapidly for 6 to 10 seconds and stop. Let the konjac sit for about 30 minutes until stable.so this is the uncooked konjac. (fig.2.3.4).
3. Place in a rice cooker, add one and a half cups of water outside the pan and steam for 30 to 40 minutes until done. Remove the steamed konjac and soak in water for 8 to 12 hours until the alkali is completely removed.so this is the cooked konjac.(fig.5.6).

生蒟蒻塊完成圖

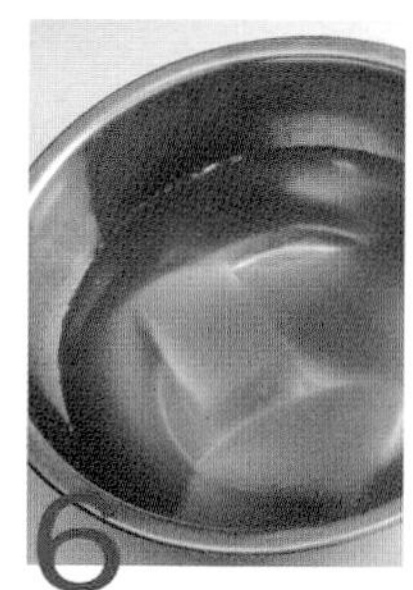

熟蒟蒻塊完成圖

蒟蒻蝦仁

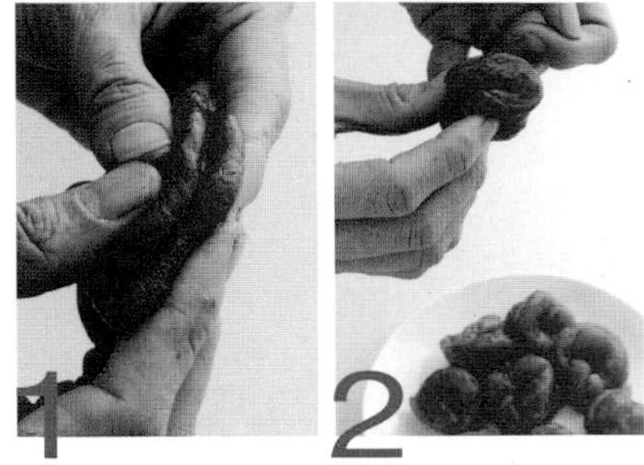

材料及器具：
蒟蒻粉40克、 鹼粉4克、胡蘿蔔丁50克、大碗、打蛋器、紗布

做法：

1. 胡蘿蔔丁加水500 c.c.放入果汁機中攪打，倒出汁液用紗布濾渣後，連同胡蘿蔔汁液加水成總量800 c.c.，倒入大碗中。
2. 參照「蒟蒻塊自己做」做法1和2使其凝成膠狀，並於室溫下放置約3~5小時至不沾手，取一小塊胡蘿蔔蒟蒻先捏成條狀，再將頭尾相連做成蝦仁狀，將捏好的蒟蒻蝦仁放入微滾的水中煮約15分鐘取出，泡冷開水去鹼。（圖1、2）

Konjac Shrimp

INGREDIENTS AND TOOLS

40g konjac powder, 4g alkali, 50g diced carrot, large mixing bowl, eggbeater, cheese cloth

METHODS

1. Beat diced carrot with 500cc of water in blender, then remove and pour the liquid through a cheesecloth to discard any dregs. The amount of carrot juice with water should be 800cc. Place mixture in a mixing bowl.
2. Follow step (1)、(2) in the P.14 "Konjac Chunks Do It Youself" and let konjac set until stable under the room temperature for 3 to 5 hours until no longer sticky to the hands. Remove a piece of carrot konjac and squeeze into a strip, then connect both ends together to make it look like a shrimp. Repeat to finish all the konjac shrimp, cook them in lightly boiling water for about 15 minutes. Remove and soak in cold water to remove the alkali. (fig1.2)

蒟蒻脆腸

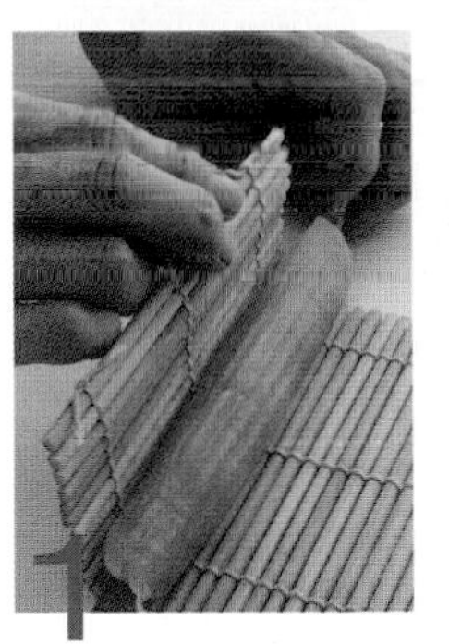

材料及器具：
生蒟蒻塊、捲壽司用竹簾、竹筷子

做法：
取一小塊生蒟蒻放在竹簾的底端上，將竹筷子放在蒟蒻塊上，拉起竹簾向前捲，使兩端接合之後連同筷子一起放入水中加熱定型。煮熟後依基本做法完成去鹼即可調理。（圖1、2）

Konjac Intestines

INGREDIENTS AND TOOLS

uncooked konjac, sushi mat, bamboo chopsticks

METHODS

Place a piece of uncooked konjac at the bottom of a sushi mat. Place the bamboo chopstick over top and roll up the sushi mat forward to make it connect together. Remove and cook in water until it is set, with the chopsticks still inside. Remove the cooked konjac intestines and follow the steps in Basic Preparation of Konjac to remove the alkali before starting to cook. (fig1.2)

蒟蒻魷魚排

材料及器具：

生蒟蒻塊、生地瓜

做法：

生地瓜去皮，取橫面切成約1公分厚的片狀，兩頭較尖的部分稍微修整。取一塊生蒟蒻包住地瓜，接縫處要捏緊。放入微滾的水中煮至半熟，取出沖涼後在沒有接縫的一面均勻間隔的劃開，放入水中繼續加熱定型。煮熟後取出地瓜，再依基本做法完成去鹼後即可調理。（圖1、2）

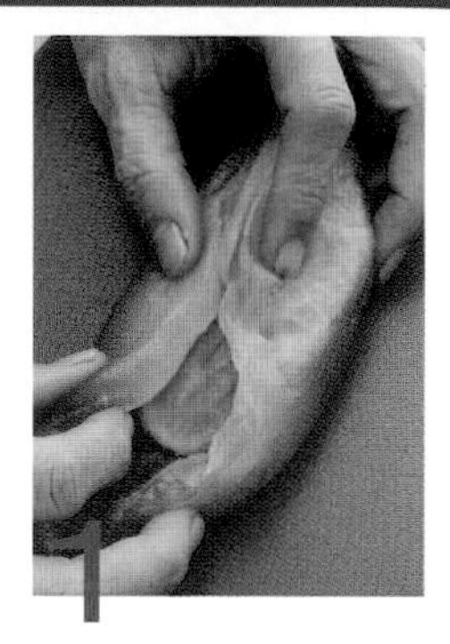

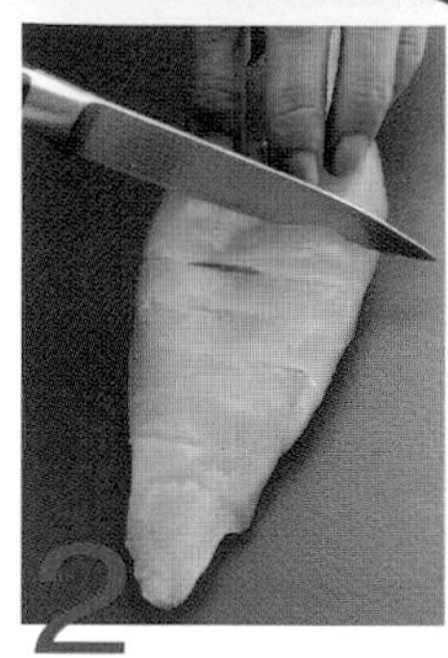

Konjac Squid Steaks

INGREDIENTS AND TOOLS

uncooked konjac, uncooked yam (fig.1)

METHODS

1. Peel yam and cut horizontally into slices 1 cm thick. Trim both pointed ends slightly. Wrap the yam slice up with a piece of uncooked konjac and seal the opening tightly. Cook in lightly boiling water until half done, remove and rinse under water to cool, then divide, not at the opening, but at the smooth part where opens evenly. Return to water and continue to cook until set. Remove and discard the yam slice, then follow the step in Basic Preparation of Konjac to remove the alkali before starting to cook. (fig1.2)

蒟蒻藕片

材料及器具：

熟蒟蒻塊、吸管、普洱茶汁

做法：

熟蒟蒻塊切成蓮藕的圓形片狀，每一片用吸管壓出環狀小洞，浸泡在潽茶汁中讓它變成天然的藕色。（圖1、2）

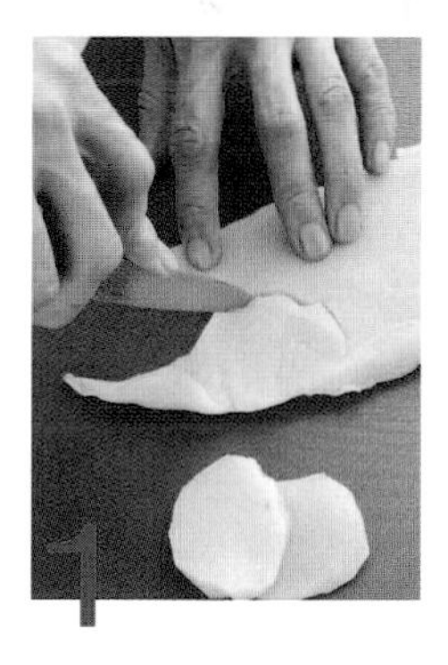

Konjac Lotus Root Slices

INGREDIENTS AND TOOLS

Cooked konjac pieces, straw, Pu-erh tea liquid

METHODS

Cut cooked konjac piece into round shape look like lotus root slices. Press with each slice with a straw into small holes. Soak in Pu-erh tea to give it a natural purple lotus root color. (fig1.2)

蒟蒻蝴蝶麵

材料及器具：

黃椒生蒟蒻塊、披薩滾刀、擀麵棍（圖1）

做法：

1.黃甜椒100克加水800c.c.打成汁，再依p.14「蒟蒻塊自己做」的做法1、2做成黃椒生蒟蒻塊。

2.取適量蒟蒻塊用擀麵棍擀平，用披薩滾刀在表面輕輕畫上一條一條的裝飾紋路，再切成5×3的長方片，每一長方片從中間捏緊做成蝴蝶狀，放入水中加熱定型。煮熟後依基本做法完成去鹼後即可調理。（圖1、2）

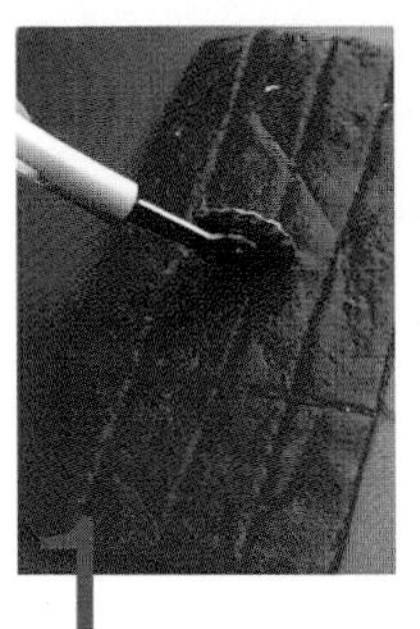

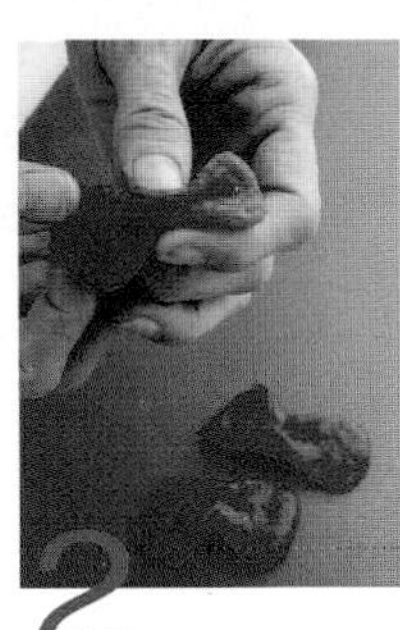

Konjac Farella

INGREDIENTS AND TOOLS

uncooked yellow pepper flavored konjac piece, pizza pastry wheel, rolling pin

METHODS

1.Blend 100g of yellow bell pepper with 800cc of water in blender to form juice, then follow the method (1) and (2) in "Konjac Chunks Do It Yourself" on p.14 to prepare yellow pepper konjac.

2.Remove suitable amount of konjac pieces and roll flat with a rolling pin. Then score with pizza wheel on the surface into serrated patterns, then cut into 5 x 3 rectangle pieces. Squeeze the center of the rectangle slightly to make it look like a butterfly, then cook in water to set. Follow the steps in the Basic Preparation of Konjac to remove the alkali before starting to cook. (fig1.2)

蒟蒻凍自己做

材料及器具：

蒟蒻凍粉40克、紅茶汁液1000 c.c.、糖100克（圖1）

做法：

將蒟蒻凍粉、糖放入紅茶汁液中攪拌均勻，以中火煮至微微沸騰後熄火，倒入耐熱模型中，於室溫下放涼再移入冰箱冰涼即可。（圖2）

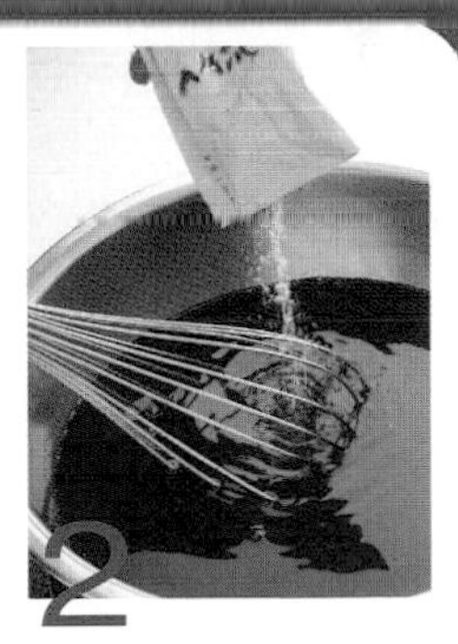

Konjac Jelly DIY

INGREDIENTS AND TOOLS

40g konjac jelly powder, 1000cc black tea liquid, 100g sugar (fig.1)

METHODS

Combine konjac jelly powder and sugar in black tea liquid well, stir until evenly-mixed, then cook on medium until lightly boiling and remove from heat. Pour into heat resistant molds and let cool under room temperature, then chill in the refrigerator. (fig.2)

KONJAC DIY Q & A
蒟蒻DIY Q&A

Q 製作蒟蒻為什麼一定要加鹼粉？

A 蒟蒻粉含有由甘露糖和葡萄糖以約3：2的β-1、4-linkage結合而成的葡甘露聚糖（Glucomannan），這是一種可逆膠體，如果與弱鹼溶液共同加熱水煮，則可形成三面體氫鍵結合的網狀結構，這種膠體網狀結構在酸與鹼的系統中有明顯的安定性，簡單的說，製作蒟蒻時加鹼粉的主要作用是將「熱可逆」（heat reversible）轉化成「熱不可逆」（heat irreversible），也就是當蒟蒻與鹼粉加熱形成「熱不可逆」的狀態之後，即使再反覆加熱也不會融化還原，這樣才能方便我們來烹調成各種食物。

Q 鹼粉是否能夠完全去除？如何節省去鹼的時間？

A 可以的，一般加在蒟蒻粉中所謂的「鹼粉」其實是統稱。例如氫氧化鈣（calcium hydroxide）、碳酸納（sodium carbonate）、碳酸氫鈉（sodium bicarbonate）等具有「鹼」特性的物質，都是可以用來與蒟蒻粉產生作用的，其中碳酸氫鈉的鹼度較低且易溶於水，在使用上較為合適。當鹼粉與蒟蒻粉經過加熱形成「熱不可逆性」（heat irreversible）之後，我們可以經由將蒟蒻塊浸泡於清水的方式讓鹼逐漸釋放於水中，達到完全不含鹼的狀態後就能放心的烹調了。

至於節省去鹼時間的方法，建議在蒟蒻蒸煮成型後，先切成小塊，再放入鍋中用清水煮沸5~10分鐘，可以加些白醋來加速去鹼的時間。當然，之後還是需要繼續將蒟蒻浸泡於乾淨的水中，直到蒟蒻變白且完全不具鹼味才可以烹調。

Q 製作蒟蒻塊時忘記放鹼粉怎麼辦？

A 蒟蒻粉和鹼粉必須充分混合才能完全發揮作用，如果在操作時忘了先放鹼粉溶於水中，就算蒟蒻粉仍可以和水凝成固態，但一經加熱後，還是會還原。最好在操作前先將所有材料都準備好，才能避免這種狀況的發生。

Q 製作蒟蒻塊時出現顆粒怎麼辦？

A 掌握以下重點就不會出現顆粒：

1.選用底面積較小的容器。
2.一定要使用打蛋器。
3.倒入蒟蒻粉之前，用打蛋器持續攪動水形成旋渦。

4.蒟蒻粉要一次全部倒入漩渦中，快速攪打6~10秒鐘。

依照以上的步驟做出來的蒟蒻塊應該是不會有顆粒的，如果還是有小部分顆粒出現，甚至有較多量的蒟蒻粉沒有被打散而形成塊狀時，只要在蒟蒻煮熟前將顆粒取出，再依正常程序操作即可，只是這種情況下做出來的蒟蒻塊，因為形成結塊的蒟蒻粉被取出而無法與水完全作用，口感會稍軟。

Q 除了用電鍋蒸熟外，還有其他方法可以製作蒟蒻嗎？

A 可以用隔水蒸煮或直接水煮的方式來將蒟蒻煮熟。隔水蒸煮是為了保持蒟蒻在尚未加熱定型之前的形狀，例如做蒟蒻生魚片時所用的方模，就可以直接放入較深的鍋中加水蒸煮，這樣做出來的蒟蒻可方便將蒟蒻切成一片一片長方形。另外，如果是做蒟蒻蝦仁、蒟蒻脆腸這類成品，可以將做好的蒟蒻成品直接放入水中煮熟。鍋中的水保持微微沸騰就好，切忌用大火，否則做出來的蒟蒻會因為過度膨脹而不好看也不好吃。

Q 為什麼做出來的蒟蒻塊會有一些坑洞或氣泡？

A 在做蒟蒻塊時，我們會用打蛋器快速攪打蒟蒻粉，所以難免會形成氣泡；不過，只要攪打的時間不超過15秒，並在停止攪打後將容器輕輕搖動或輕敲，就可以避免形成氣泡過多的現象。

Q 如何變化出蒟蒻的各種顏色及形狀？

A 蒟蒻粉本身並不具任何顏色，可以利用許多食物的天然色素來豐富它的顏色，比如胡蘿蔔、甜菜根、綠色葉菜、各色甜椒、黑木耳、黑芝麻粉等都是很好的材料。在製作有色蒟蒻時，先確定想要的Q度口感，假設用1份（40g）的蒟蒻粉對800c.c.的液體，這800 c.c.的液體中就應該包含大約200 c.c.的胡蘿蔔汁或蔬菜汁（有色汁與液體約為1：3），再按基本步驟操作即可。如果是用粉類來做顏色的變化，就直接將各種粉加進水中就可以了。

蒟蒻塊在尚未加熱形成固體前就像黏土一樣，可以做出很多不同的形狀；但要注意的是必須讓蒟蒻塊在室溫下至水分蒸發到不會沾手才能開始操作，至於水分收乾的時間則須視當時的溫度及濕度而定，當氣溫高、濕度低時大約需要1~2小時，如果溫度低、濕度高時，可能需要6~8小時。總之，可以不時抓起一小塊蒟蒻來試一下，多做幾次有經驗後就可以瞭解最佳時機。

Q 蒟蒻的口感可以自行控制嗎？

A 可以，蒟蒻粉和蒟蒻凍粉都可依自己的喜好來調整Q度。蒟蒻粉和水量的比例是影響口感的關鍵，喜歡有咬勁、Q度高的人可以少放些水；相反的，如果家裡有小朋友或牙口不好的老年人，就要多放些水讓蒟蒻吃起來較不費勁。在基本做法裡，示範的比例是20倍，也就是蒟蒻粉：水＝40克：800c.c.，如果要增減水分，可以在100 c.c.~150 c.c.間做調整。蒟蒻凍粉的Q度除了受水分影響外，「酸度」也是影響Q度的因素。以果汁蒟蒻凍為例，相同分量的檸檬汁和蘋果汁做出來的蒟蒻果凍，檸檬凍的口感就會軟一些。每個人對Q度的感覺不盡相同，可以多做幾次，就能找出自己最喜歡的口感。

Q 蒟蒻粉和蒟蒻凍粉有何不同？可以互相代替嗎？

A 蒟蒻粉和蒟蒻凍粉的成分不同，用法也不同，是不可以互相替代的。蒟蒻粉是100%的純粉，當蒟蒻純粉與鹼粉經過加熱形成熱不可逆的作用後，可以反覆加熱而不會還原，因此適合各種烹調方式，例如炒、煮、燉、炸等。蒟蒻凍粉的成分除了蒟蒻粉之外，還調配了適當比例的海藻膠；海藻膠的保水性佳而蒟蒻膠的Q度高，在結合了這兩種膠體的特性後，做成的蒟蒻凍會有特別的口感。但基本上蒟蒻凍是不可以加熱的，加熱之後的蒟蒻凍會還原溶解。所謂「不可加熱」，是指蒟蒻凍體本身不能持續高溫加熱，但蒟蒻凍外的液體則不在此限，這本食譜中的「莧菜豆腐腦兒」，就是蒟蒻凍熱食的做法。

Q 蒟蒻製成品可以存放在冷凍庫嗎?

A 純蒟蒻製成品只能放在冰箱冷藏保存，不能放進冷凍庫。蒟蒻製成品幾乎有90％都是水分，一旦放入冷凍庫後，會產生冰晶現象破壞網狀組織，當蒟蒻在室溫下退冰後水分被釋出，失去彈性的蒟蒻製成品吃起來就會像橡皮筋一樣很難咀嚼。市面上有許多標示含有蒟蒻成分的製成品，因為添加了大部分的修飾澱粉、大豆蛋白等做為抗凍劑，所以能夠冷凍保存，但相對於純蒟蒻製品，這類的製成品中所含的蒟蒻成分低了很多。

Q Why does the alkali have to be added when making konjac?

A Konjac powder is a β -1 and 4 linked combination of glucomannan with manose and glucose in a 3:2 proportion. If combined with a weak alkali solution and heated in hot water, it forms a trihedral hydrogen bond. This kind of structure in an acid and alkali system has an obvious stability. Simply speaking, the main function of adding alkali when making konjac, is to turn something that is heat reversible into something that is heat irreversible. That is, once the konjac and alkali have been heated, a heat irreversible state is formed. It will never dissolve even after repeated heating. Thus we can use it in cooking.

Q Can the alkali be removed completely? How can we save time in the removal process?

A Yes, the alkali powder added to the konjac is only a common name for a number of substances, such as calcium hydroxide, sodium carbonate, and sodium bicarbonate, materials that have "alkali" characteristics. Any of these can be used to make konjac powder. Among them, sodium bicarbonate contains the lowest alkali level and is easy dissolved in water, making it perfect for konjac. After the alkali powder and konjac powder become heat irreversible when heating, we can soak the konjac jelly pieces in water to allow the alkali to be released into water until a state of no alkali has been reached. Then we can use it in cooking with ease. To save time, I suggest that after the konjac is steamed and set, cut it into small pieces first, then boil in water for 5 to 10 minutes. Adding a little white vinegar will speed up the process. Of course, konjac still needs to be soaked in clean water until it is completely white and no alkali is left.

Q What should we do if we forgot the alkali powder when making the konjac pieces?

A Konjac powder and alkali powder have to be completely combined for best results. If alkali powder is not dissolved in water during the operation, the konjac powder can still be combined with water and formed into jelly. However, if heated again, it will dissolve again. Therefore, the best way is to ready all the ingredients, to prevent this kind of situation.

Q What should we do if there lumps appear during preparation of the konjac pieces?

A If you grasp these important tips, there will not be any lumps:

* Use molds with small bottom area.
* Use an eggbeater.
* Beat the water continuously to form whirlpools before adding the konjac pow-

der.

* Add konjac powder a little at a time to the swirling water and beat rapidly for 6 to 10 seconds.

If you follow the steps above exactly, there should not be any lumps in the konjac pieces. If lumps appear, or large lumps appear due to improper beating, remove crumbs before steaming. Then follow the procedures and do it again. However, in this situation, because some of the konjac pieces have been removed and will not be well-combined with water, the texture will be softer.

Q Are there any other ways of preparing konjac jelly besides steaming in the rice cooker?

A Konjac jelly can be prepared by either steaming in the double-boiler or cooking directly in water. Using a double-boiler to steam konjac can preserve its shape before steaming. For example, the konjac sashimi square mold can be steamed directly in the double boiler with konjac inside, then the konjac jelly can be conveniently cut into long rectangular slices after steaming. Others recipes, like konjac shrimp and konjac intestines are made separately and can be cooked directly in water. Just keep the water boiling lightly, and do not cook over high heat, or the konjac will be ugly and not very tasty due to over-expansion.

Q Why are there holes or air bubbles in the konjac jelly after it has been cooked?

A During the process of making konjac jelly, we use an eggbeater to beat the konjac powder at high speed, so it is inevitable that bubbles will form. However, if we beat it less than 15 seconds, then stop beating and shake or whisk the mixer gently, bubble formation will be reduced.

Q How is the konjac jelly prepared into different colors and shapes?

Konjac powder itself is colorless. Many natural food colors from food can be used to color it, like carrots, sweet beets, green vegetable leaves, colored bell peppers, black woodear fungus, black sesame seed powder, and so forth. When trying to prepare colored konjac jelly, first make sure the konjac is chewy. For example, 1 portion of konjac powder (40g) with 800cc of liquid should contain about 200cc of carrot juice or vegetable juice (the ratio of coloring juice and liquid should be about 1:3. If powder is used to vary the color, then add powder directly to the water.

Before setting, konjac jelly is just like clay and can be made into different shapes. One important thing is to let the konjac jelly sit at room temperature until the liquid has evaporated and it is easy to handle. The evaporation time depends on the temperature and humidity. When the temperature is high and humidity is low, it takes 1 to 2 hours. If the temperature is low and humidity is high, it might take 6 to 8 hours. You can test it by tasting a small amount of it. As practice makes perfect, sooner or later you will understand when it is done.

Q Can we control the texture of the konjac?

A Yes, konjac powder and konjac jelly powder can be adjusted to fit your preferred level of chewiness. The proportion of konjac powder and water is the key to the texture. If you prefer chewy texture, add less water. If there are kids or any old people in the family, then add more water to make the konjac softer. In the Basic Methods of Preparing Konjac, the demonstrated proportion is 1:20, that is, 40g of konjac powder: 800cc of water. If you increase the water, you can vary it by 100cc to 150cc. In addition to water, which influences the chewiness of konjac jelly powder, the "sourness" is also a factor. Let's take juice konjac as an example. If identical amounts of lemon juice and apple juice are used to make konjac jelly, the lemon konjac jelly will be softer. Everyone has their own preferred level of chewiness, so make it more often to find out what your favorite texture is.

Q What is the difference between konjac powder and konjac jelly powder? Can they substitute for one another?

A Both the ingredients of konjac powder and its uses are different from konjac jelly, so they cannot substitute for one another. Konjac powder is 100% pure powder, which, after heating with alkali, is suitable for any cooking method. Konjac jelly powder is konjac powder with a suitable proportion of seaweed gum. Seaweed gum retains water, and chewiness of konjac gum is high. The combination of these two characteristics gives the konjac jelly a special texture. Basically, konjac jelly cannot be heated, or it will dissolve. "No heating allowed" means that the konjac jelly cannot be continuously heated on high, though you can add any liquid you want to it. The Konjac Tofu with Amaranth dish in this cookbook is a hot dish that uses konjac jelly.

Q Can konjac jelly products be stored in the freezer?

A Pure konjac jelly product can only be stored in the refrigerator, not in the freezer. Konjac jelly products are 90% water. If frozen, ice crystals form and the structure breaks down. When the water is released once it is defrosted at room temperature, the konjac jelly will taste awful, like a rubber band. There are many konjac products sold in the market and most of them have added starch or bean protein so that they can tolerate freezing. That is the reason why they can be put in the freezer. As for pure konjac products, those sold in the market all contain very low konjac.

想要自己做蒟蒻食品，可以到以下商店購買蒟蒻粉和蒟蒻凍粉：

地區	商店	電話	地址
大安區	塘塘廚坊	(02)2732-1243	樂利路42巷2號
	老農夫	(02)2736-9825	安和路2段171巷2號1樓
	綠色小鎮台大店	(02)2369-6206	羅斯福路283巷38號
信義區	有機田生活坊	(02)2762-7540	松山路336巷15號(虎林街126號對面)
	綠色小鎮東門店	(02)2327-8066	臨沂街57號
	綠色小鎮北醫加盟店	(02)2737-0838	吳興街220巷23號
中山區	佛法山有機蔬菜廣場	(02)2506-0346	松江路123巷5號1樓
	老農小舖	(02)2563-7884	松江路330巷67號1樓
	大自然馬偕	(02)2567-7216	中山北路2段96巷1號
	無毒之家大直店	(02)2532-8785	北安路584號1樓
	義興烘焙	(02)2760-8115	富錦街578號(民權大橋下)
士林區	樸園忠誠店	(02)2833-6258	忠誠路2段50巷9號1樓
	樸園石牌店	(02)2828-7336	石牌路2段122號
	飛訊烘焙	(02)2883-0000	承德路4段277巷83號
內湖區	皇品烘焙原料	(02)2797-3932	內湖路2段13號
基隆市	富盛烘焙材料行	(02)2425-9255	基隆市南榮路64巷8號2樓
板橋市	綠色小鎮板橋店	(02)2258-8880	板橋市松柏街57號
新店市	樸園(健生店)	(02)2911-3205	新店市大豐路26號(大坪林站麥當勞旁巷子)
中和市	若霖	(02)2247-3737	中和市復興路297號
	若霖	(02)2249-5958	中和市南山路67號
新莊市	香園有機小站	(02)8994-4006	新莊市明中街12-29號
	綠色小鎮 新莊店	(02)8991-2516	新莊市建新街11號
	小蕃茄有機坊	(02)2204-3636	新莊市四維路210-5號
樹林市	光緣有機樹林店	(02)2681-4122	樹林市育英街67號
土城市	宜欣有機	(02)2265-9367	土城市中央路1段233號
桃園市	蕃薯藤	(03)333-4546	桃園市縣府路130號
	天然生機養生館	(03)358-2964	桃園市大業路1段290號
中壢市	綠色奇蹟	(03)402-6346	中壢市中央路2段16號
平鎮市	振宇商行	(03)459-3888	平鎮市金陵路2段3號
台中市	有機世界	(04)2243-9195	台中市崇德2路2段7號
	綠色小鎮台中店	(04)2302-3381	台中市美村路1段359號
	三贏人生(澄清店)	(04)2463-8966	台中市中港路3段116-5號
台中縣	三贏人生(菩提店)	(04)2486-1515	台中縣大里市中興路2段615-1號
彰化市	三贏人生(彰基店)	(047)258-334	彰化市中興路158號
嘉義市	福美珍烘焙原料店	(05)222-4824	嘉義市西榮街135號
台南市	清涼世界興華店	(06)228-9896	台南市興華街7號
	清涼世界西門店	(06)282-7375	台南市西門路4段283號
台南縣	大地生機飲食推廣中心	(06)633-7686	台南縣新營市三民路190-2號
高雄縣	綠色生活	(07)711-4745	高雄縣鳳山市南昌街76巷23號
	清涼世界岡山店	(07)621-8617	高雄縣岡山鎮壽天路20號

Jelly* Dessert 果凍・點心

薑汁蕃茄蒟蒻凍Tomato Konjac Jelly with Ginger Sauce

金黃奶凍Golden Milk Konjac

乾果蒟蒻凍Dried Fruit Konjac Jelly

蜜豆凍Konjac Jelly with Sugar Snap Peas and Lotus Seeds

翠綠相思凍Young Soybeans Konjac Jelly with Aduki Beans

黑芝麻凍Black Sesame Konjac Jelly

藍莓醋凍Konjac Jelly in Blueberry Vinegar

小麥草凍Wheat Grass Konjac Jelly with Molasses

黑磚奶茶Konjac Milk Tea

蕃茄蒟蒻凍飲Konjac Cubes in Tomato and Plum Juice

黑糖水晶凍凍條Konjac Noodles with Dark Brown Sugar

蜜漬黑豆凍Candied Black Soybeans Konjac Jelly

水晶紫米丸子Crystal Purple Rice Balls

草莓大福Ta-Fu Strawberry Konjac with Aduki Beans

蒟蒻香蕉卷Banana in Konjac Rolls

蒟蒻點心條Baked Konjac Snack Bars

Tomato Konjac Jelly with Ginger Sauce

薑汁蕃茄蒟蒻凍

材料：（3~4人份）
蒟蒻凍粉40克、蕃茄汁1000c.c.

醬汁：
嫩薑末1小匙、素蠔油1大匙、冷開水1大匙

做法：
1.蒟蒻凍粉倒入蕃茄汁中攪拌均勻，以中火煮至微微沸騰即熄火。
2.將做法1.倒入耐熱模型中，待涼切塊即可沾醬食用。

INGREDIENTS:

3-4 people
40g konjac powder, 1000cc tomato juice

DIPPING SAUCE:

1t minced tender ginger, 1T vegetarian oyster sauce, 1T cold water

METHODS:

1.Combine tomato juice well with konjac powder, stir until evenly mixed and heat on medium until lightly boiling, then remove from heat immediately.
2.Pour method (1) into heat-resistant molds until set and cold. Remove and cut into serving sizes. Serve with the dipping sauce on the side as a dip.

1.紅咚咚的蕃茄凍，單吃已經非常可口，再沾上美味醬汁，怎麼停得了？
2.嫩薑末搭配素蠔油調成的醬汁，除了可以沾蕃茄凍，還可以沾生蕃茄食用，味道相當特別！

1.Red color tomato konjac jelly is very pleasing to the eye. By itself it is delicious enough, never mind the delicious dipping sauce. How can anyone stop?
2.Sauce with minced tender ginger and vegetarian oyster sauce can not only be a dip for the konjac jelly, but is also good with the tomatoes. The flavor is unique.

Colden Milk Konjac

金黃奶凍

材料：（6~8人份）
蒟蒻凍粉40克、南瓜300克、水700c.c.、鮮奶200c.c.、鮮奶油300克

做法：
1.南瓜去籽切塊蒸熟，加水以果汁機打成汁。
2.將南瓜汁、鮮奶、蒟蒻凍粉一起攪拌均勻，以中火煮沸後熄火。
3.加入鮮奶油，用打蛋器仔細拌勻後倒入模型，降溫後移入冰箱冷藏。

INGREDIENTS:
6-8 people
40g konjac powder, 300g pumpkin, 700cc water, 200cc milk, 300g whipping cream

METHODS:
1. Discard seeds from pumpkin and cut into pieces, then steam until done. Mix well with water in blender and blend to make into pumpkin liquid.
2. Combine pumpkin liquid, milk and konjac powder well together. Bring to boil over medium heat, then remove from heat.
3. Add whipping cream to mix. Beat with an egg beater until evenly mixed, then pour into molds to cool first. Remove to chill in refrigerator.

1.南瓜富含維生素A及B₁，具護眼功效，也可保養皮膚；金黃色的顏色，也讓人在視覺上大飽眼福。
2.煮南瓜湯時，取出的熟南瓜籽不要丟棄，可以像嗑瓜子般當成零食吃。

1. Pumpkin contains vitamin A and B_1. It protects the eyes and skin. The golden dish is also visually satisfying for diners.
2. Do not discard the cooked pumpkin seeds, they can be served as snack like watermelon seeds.

Dried Fruit Konjac Jelly

乾果蒟蒻凍

材料：（6~8人份）
蒟蒻凍粉40克、冷開水1000c.c.、各類乾果200克、糖100克

做法：
1.蒟蒻凍粉加水攪拌均勻，以中火煮至約60℃（140℉）時放入乾果及糖，續煮至微微沸騰即熄火。
2.將做法1.倒入耐熱模型中，在室溫下放涼後移入冰箱冰40~60分鐘即可。

INGREDIENTS:
6-8 people
40g konjac powder, 1000cc cold water, 200g all kinds of dried fruit, 100g sugar

METHODS:
1. Combine konjac powder with water and cook over medium heat until the temperature reaches 60℃（140℉）. Add dried fruit and sugar. Continue heating until lightly boiling and remove from heat.
2. Pour method (1) into a heat resistant mold. Let cool at room temperature, then remove to chill in the refrigerator for about 40 to 60 minutes until set. Serve.

選用的乾果如果是像杏子這類較大型的，要先用冷開水泡軟才不至於過硬。
各類乾果是很營養的零食，搭配蒟蒻做成果凍就更健康了。

Large size dried fruit, such as apricot, has to be soaked in cold water until soft to prevent it from being too hard to serve. Dried fruit is a very nutritious snack and is even healthier made into jelly with konjac.

Konjac Jelly with Sugar Snap Peas and Lotus Seeds

蜜豆凍

材料：（6~8人份）
蒟蒻凍粉40克、糖漬蜜豆100克、熟地瓜100克、煮熟小湯圓少許、 黑糖80克、煮熟雪蓮子50克、水1200c.c.

做法：
1.蒟蒻凍粉加入水中均勻攪拌，用中火煮至約60°C（140°F）時放入黑糖及蜜豆續煮。
2.沸騰後即可熄火，加入各式配料，趁熱食用即可。

INGREDIENTS:

6-8 people
40g konjac powder, 100g candied sugar snap peas, 100g cooked yam, small cooked soup dumplings as needed, 80g dark brown sugar, 50g cooked lotus seeds, 1200cc water

METHODS:

1. Mix konjac powder with water and stir until evenly mixed. Cook over medium heat until the temperature reaches 60°C（140°F）, add dark brown sugar and candied sugar snap peas to mix, then continue cooking
2. Bring to a boil, remove from heat and add all the remaining ingredients. Serve when it is still steaming.

1.趁熱食用的蒟蒻凍很像吃燒仙草，不過卻少了澱粉，多了纖維。
2.蒟蒻凍多為涼食，這種另類吃法，你一定要試試看！
3.雪蓮子是一種口感近似栗子的豆類，可以在超市買得到。

1. The steaming konjac jelly tastes like Hsian Tsao solution, yet without starch and more fiber.
2. Konjac jelly is usually served cold. Dishes like this one are very special. You've got to try it!
3. Snow lotus seed is a kind of bean, with a texture similar to chestnuts, that can be found in the supermarket.

Young Soybeans Konjac Jelly with Aduki Beans

蒟蒻凍粉40克、水1000 c.c.、毛豆仁300克、蜜漬紅豆200克、糖70克、鮮奶油適量

做法：

1. 毛豆仁去膜洗淨，放入沸水中煮5分鐘取出。
2. 煮熟的毛豆仁加水，放入果汁機中打碎後倒入鍋中。
3. 續加入蒟蒻凍粉、糖均勻攪拌，開中火慢慢煮沸。
4. 將做法3.倒入模型中，於室溫下放涼，移入冰箱冰約50分鐘後，倒扣在盤中，加些蜜漬紅豆，淋上鮮奶油即可。

INGREDIENTS:

8-10 people

40g konjac powder, 1000cc water, 300g young soybeans, 200g candied sugared aduki beans, 70g sugar, whipping cream as needed

METHODS:

1. Remove shells from young soybeans and rinse well, then cook in boiling water for 5 minutes and remove.
2. Add water to cooked young soybeans and blend in blender until well-mashed, then remove to cooking pan.
3. Add konjac powder and sugar to mashed young soybeans in pan, stir until evenly mixed and bring to boil over medium.
4. Pour method (3) into molds and let sit under the room temperature until cool. Remove to chill in the refrigerator for about 50 minutes until set. Remove and flip over onto the serving plate. Add some candied sugared aduki beans and drizzle with whipping cream. Serve.

1. 毛豆必須先煮過才能去除豆腥味。
2. 毛豆含有極豐富的植物蛋白，為了保存養分不需濾渣，所以儘可能打得細一些，才不致影響口感。

1. Young soybeans have to be cooked first to remove the fishy bean flavor.
2. Young soybeans are high in vegetarian plant protein. In order to retain the nutrition, it is not necessary to discard the dregs. Blend as finely as possible so that the texture will not be affected.

Black Sesame Konjac Jelly

材料：（6~8人份）
蒟蒻凍粉40克、黑芝麻粉5大匙、糖2大匙、冷開水1000c.c.

做法：

1.冷開水中放入蒟蒻凍粉攪拌均勻，以中火慢慢煮。
2.煮至約80°C（176°F）時，加入糖及黑芝麻粉，續煮至沸騰熄火。
3.將做法2.倒入耐熱模型中，在室溫下放涼後移入冰箱即可。

INGREDIENTS:

6-8 people
40g konjac powder, 5T black sesame powder, 2T sugar, 1000cc cold water

METHODS:

1.Combine konjac powder with cold water until evenly mixed, then cook over medium heat slowly.
2.Continue cooking until the temperature reaches 80°C（176°F）, add sugar and black sesame powder to mix. Continue until boiling and remove from heat.
3.Pour method (2) into heat-resistant molds and let cool at room temperature before removing to the refrigerator.

1.如果選用市售的含糖芝麻粉就可不必再加糖了。常吃黑芝麻可使皮膚有光澤、頭髮烏黑。
2.黑芝麻含有維生素A、維生素B_1、維生素B_2、維生素E、維生素F及鈣、鐵等多種營養素，對於胃酸過多、胃擴張等亦有功效。

1.If the sweetened black sesame powder sold in market is used, there is no need to add more sugar. Black sesame seeds help the skin and hair shine.
2.Black sesame seeds contain many minerals, such as vitamin A、B_1、B_2、E and F, along with calcium, and iron. It also has the effect of curing stomach acid and swollen stomach.

Konjac Jelly in Blueberry Vinegar

藍莓醋凍

材料：（3~4人份）
蒟蒻凍粉40克、藍莓醋30c.c.、糖80克、水850C.C.

做法：
1.蒟蒻凍粉加水、糖均勻攪拌後，以中火煮至微微沸騰熄火。
2.當溫度降至80°C（176°F）時，將藍莓醋倒入均勻攪拌即可。

INGREDIENTS:
3-4 people
40g konjac powder, 30cc blueberry vinegar, 80g sugar, 850cc water

METHODS:
1.Combine konjac powder and sugar in water, stir until evenly mixed and cook over medium heat until lightly boiling. Remove from heat.
2.When the temperature drops to 80°C (176°F), pour blueberry vinegar gently and stir until evenly mixed.

1.相信大家都知道「多喝醋、好處多」，藍莓富含「花青素」，對活化血管、抗氧化、保護眼睛有相當的助益，為了保存營養素，最後再加入藍莓醋就可以了。
2.液體的酸度較高時會降低蒟蒻的Q度，所以，同分量的蒟蒻凍粉在搭配高酸度的汁液時，水分要略減少，才能維持QQ的口感。

1.I believe that everybody already knows the saying "more vinegar, more benefit." Blueberry contains anthocyanidins, which benefit the blood vessels, have anti-oxidant properties, and protect the eyes. In order to preserve its nutrition, add it at the end.
2.The sourness of the liquid will decrease the chewiness of the konjac, so decrease the water to maintain the chewiness of the konjac.

Wheat Grass Konjac Jelly with Molasses

小麥草凍

材料：（6~8人份）
蒟蒻凍粉40克、水1000c.c.、小麥草汁200c.c.、黑糖蜜適量

做法：

1. 蒟蒻凍粉加水攪拌均勻，以中火煮沸後熄火，在室溫下放涼降溫。
2. 當蒟蒻凍汁液的溫度大約降至85°C（185°F）時，加入小麥草汁再度攪拌均勻。
3. 將做法2.倒入模型中，待完全降溫後移入冰箱冰約60分鐘，吃時再淋上黑糖蜜即可。

INGREDIENTS:

6-8 people
40g konjac powder, 1000cc water, 200cc wheat grass juice, molasses as needed

METHODS:

1. Combine konjac powder with water and stir until evenly mixed. Bring to boil over medium heat, then remove from heat and let cool at room temperature.
2. When the temperature of the konjac liquid is about 85°C（185°F）, then pour in wheat grass juice and mix until evenly mixed.
3. Pour method (2) into molds and let sit until completely cooled. Remove to the refrigerator and chill for about 60 minutes until set. Drizzle with molasses before serving.

1. 小麥草汁這類應避免高溫破壞營養素的液體，都可以上述方式來製作蒟蒻凍。
2. 若在一般市場買不到小麥草汁，可到生機飲食店購買，而且必須即時操作，否則會氧化變成黑褐色。

TIPS

1. To prevent the nutrition of the wheat grass from being destroyed by the high temperature, use the method shown above to make the konjac jelly.
2. If wheat grass juice is not available at the traditional market, purchase it at the organic food and drink supply store. Use immediately after purchasing, or it will oxidize and turn dark brown.

Konjac Milk Tea

黑磚奶茶

材料：（4~5人份）
蒟蒻凍粉40克、黑糖80克、紅茶10克、冷水800c.c.、熱開水1300c.c.、鮮奶200c.c.、糖50克

做法：
1.將蒟蒻凍粉、黑糖加入800c.c.的冷水攪拌均勻後煮沸，倒入淺盤中待涼切成小丁。
2.熱開水加紅茶、糖一起煮沸，瀝去茶渣待涼後放進冰箱。
3.每杯250c.c.的紅茶放入約100克的蒟蒻黑糖塊及適量的鮮奶即可。

INGREDIENTS:

4-5 people
40g konjac powder, 80g dark brown sugar, 10g black tea, 800cc cold water, 1300cc hot water, 200cc milk, 50g sugar

METHODS:
1. Combine konjac powder and dark brown sugar in 800cc of cold water, stir until evenly mixed and bring to a boil in pan. Pour over to a shallow plate to cool and set, then cut into small dices.
2. Add black tea and sugar to hot water, bring to a boil and discard tea leaves and let cool first before chilling in the refrigerator.
3. Pour 250cc of black tea in each cup with 100g of dark brown sugar konjac along with a suitable amount of milk added. Serve.

1.市售的飲料也可以在家自己做，除了省錢之外，「衛生」看得到。如果將蒟蒻塊切得小一些，可用吸管來喝，咕嚕咕嚕的……感覺很特別。
2.黑糖含豐富的鐵質，可有效補血，對女性是一大補品。另還有散瘀、暖肝、去寒等功效，對健康有益。

1. Drinks sold in the market can be prepared at home to save money. It is also cleaner and safer. Cut the konjac pieces into smaller pieces and suck up with a straw. You will have a special satisfied feeling.
2. Dark brown sugar is high in iron. It supplements the blood and is a great supplement for women. It also has the effect of curing thick blood due to menses, warming the liver and releasing chill. It is really good for the body.

Konjac Cubes in Tomato and Plum Juice

材料：（4~5人份）
蒟蒻凍粉40克、蕃茄汁1000c.c.、梅漿200克、冷開水700c.c.

做法：
1.冷開中倒入蒟蒻凍粉攪拌均勻，以中火煮至沸騰後熄火，倒入耐熱模型中，待涼移入冰箱。
2.將蒟蒻凍切成小丁放入蕃茄汁中，每杯中放1大匙梅漿，攪拌均勻後即可飲用。

INGREDIENTS:

4-5 people
40g konjac powder, 1000cc tomato juice, 200g concentrated plum juice, 700cc cold water

METHODS:

1. Mix konjac powder well in water and stir until evenly mixed. Bring to boil over medium heat, then remove from heat. Pour into heat-resistant molds and let cool before chilling in the refrigerator.
2. Cut konjac jelly into small cubes and place inside the glasses with the tomato juice. Add 1T of concentrated plum juice to each glass and stir until evenly-mixed. Serve.

1.這是夏日午後的最佳清涼養生飲品，可以將做好的蒟蒻凍塊放進保鮮盒中冷藏，每次只切適當的量來吃就可以了。
2.挑選蕃茄時，以顏色愈紅的愈好，因其所含茄紅素較多，營養素也較完整。

1. This is a best cool light nourishing drink for the hot summer days. The konjac jelly can be stored in the well-sealed containers in the refrigerator. Remove a little when needed.
2. Selecting tomatoes that are as red as possible because they contain more lycopene and have more complete nutrition.

Konjac Noodles with Dark Brown Sugar

黑糖水晶凍凍條

材料： （5~6人份）
蒟蒻凍粉40克、水1000c.c.

醬汁：
黑糖300克、水150c.c.

做法：
1.蒟蒻凍粉加水1000c.c.攪拌均勻，以中火煮沸後熄火，倒入耐熱淺盤中，待涼放入冰箱30分鐘後取出切成細條狀。
2.將黑糖及水用小火熬煮成濃稠狀後放涼。
3.將水晶蒟蒻麵盛入碗中，淋上黑糖醬汁即可。

INGREDIENTS:
5-6 people
40g konjac powder, 1000cc water

SUGAR SAUCE:
300g dark brown sugar, 150cc water

METHODS:
1. Combine konjac powder with 1000cc of water until evenly mixed. Bring to a boil over medium heat and remove from heat. Pour into heat resistant shallow plate to cool, then remove to chill in the refrigerator for 30 minutes.` Remove and cut into fine strips to make noodles.
2. Cook dark brown sugar and water over low heat until thickened, then let cool.
3. Transfer konjac noodles to serving bowl and drizzle with dark brown sugar sauce. Serve.

1.切成細條狀的蒟蒻凍模樣晶瑩剔透，深受大家喜愛，搭配濃郁的黑糖汁，一定會成為炎炎夏日裡最受小朋友歡迎的點心。
2.黑糖的純度不及一般白糖和砂糖，每100克的熱量約為50卡，較一般糖低，加上風味特別，當吃膩普通糖類時，不妨換個口味，嘗嘗黑糖的美味。

1. Konjac cut into fine strips is transparent like crystal, and is popular with everyone. With dark brown sugar sauce, it becomes a popular dessert for small kids on hot summer days.
2. Dark brown sugar is not as refined as ordinary white sugar or granulated sugar. Every 100g has about 50 calories, lower than ordinary sugar. When you are tired of ordinary sugar, why not try this dark brown sugar with its special and delicious flavor.

Candied Black Soybeans Konjac Jelly

蜜漬黑豆凍

材料：（8~10人份）
蒟蒻凍粉40克、水1200c.c.、蜜漬黑豆200克、黑糖50克

做法：
1.蒟蒻凍粉加水、黑糖攪拌均勻，以中火煮至約60℃（140℉）時加入黑豆續煮至沸騰。
2.將做法1.倒入耐熱模型中，在室溫下放涼後，移入冰箱冰50~60分鐘即可。

INGREDIENTS:

8-10 people
40g konjac powder, 1200cc water, 200g candied black soybeans, 50g dark brown sugar

METHODS:

1. Combine konjac powder with water and dark brown sugar well in bowl, stir until evenly mixed and cook over medium heat until the temperature reaches 60℃（140℉）, then add black soybeans to mix. Continue heating until boiling.
2. Add method (1) to heat resistant molds and let cool until the room temperature first. Remove to chill in the refrigerator for about 50 to 60 minutes until set. Remove and serve.

1.黑豆是一種滋補的養生食品，經常食用對健康有益。如果會自釀黑豆酒，那麼就可以利用釀酒後的黑豆來做這道滋補的健康點心。
2.若嫌自製蜜漬黑豆麻煩，可在較大的超市或百貨公司買到。

1. Black soybeans are a kind of nourishing and supplementing food, beneficial to your bodily health if served more often. If you brew black soybean wine at home, the left-over black soybeans can be used to prepare this delicious healthy dessert.
2. If you think candied sugared black soybeans are too much trouble to make, purchase them in large supermarkets or department stores.

Crystal Purple Rice Balls

Crystal Purple 水晶紫米丸子 Rice Balls

材料：（3~4人份）
蒟蒻凍粉20克、紫糯米200克、圓糯米100克、桂圓肉50克、核桃50克

做法：
1.蒟蒻凍粉加水煮沸，倒入淺盤放涼後切成每片約10公分正方的薄片。
2.菠菜去葉留梗燙熟，備用。
3.兩種糯米洗淨，加360c.c.的水浸泡3小時後蒸熟取出，趁熱拌入桂圓肉、核桃、糖，待涼後做成每個約80克的圓形丸子。
4.將糯米球放在蒟蒻片中，拉起四個角，再用菠菜梗綁起來即可。

INGREDIENTS: 3-4 people
20g konjac powder, 200g purple rice, 100g round sticky rice, 50g longan flesh, 50g walnut, 500g water, 50g sugar

METHODS:
1. Combine konjac powder with water and bring to a boil. Remove and let cool in a shallow plate, then cut into square slices about 10cm in size.
2. Discard leaves and retain the stems from spinach, blanch in boiling water until done and remove.
3. Rinse both kinds of rice well, soak in 360cc of water for 3 hours, then steam until done and remove. Blend in longan flesh, walnut and sugar when the rice is still steaming. Divide into small equal portions about 80g and roll each into a round ball.
4. Place each rice ball in the center of each konjac square slice. Pull up the four corners and tie well with spinach stem. Serve.

1.水晶紫米丸子在剔透的蒟蒻薄片中造型很討喜，製作時需多花一些時間，但卻是三五好友相聚時最適合的茶點。
2.糯米較不易消化，不可一次吃太多，否則容易消化不良。
3.桂圓又稱龍眼，具良好的滋補作用，尤其可用在婦女產後調理及病後體力衰弱。

1. The konjac slice with purple rice ball inside looks very attractive. It takes some time to prepare, but it is perfect for friends getting together for tea.
2. Sticky rice is not easy to digest, so do not eat too much at a time, or the stomach might not be able to handle it.
3. Longan flesh is nourishing, and is used in caring for women after birth and for weakness after sickness.

Ta-Fu Strawberry Konjac with Aduki Beans

Ta-Fu Strawberry Konjac

草莓大福

with Aduki Beans

材料： （8~10人份）
蒟蒻凍粉20克、水400c.c.、糯米粉200克、水200c.c.、草莓10顆、紅豆餡100克、太白粉3大匙

做法：

1. 糯米粉加水200c.c.拌勻後倒入淺盤中，放入蒸籠蒸10分鐘，取出放涼。
2. 蒟蒻凍粉加水400c.c.煮成蒟蒻凍汁，倒入淺盤中做成薄薄的一張蒟蒻凍片，待涼後再切成四方形的10等份。
3. 將紅豆餡分成10等份，1份包住一顆草莓。
4. 糯米片放在淺盤中，均勻撒上太白粉後再分成10等份的薄片，每一份先放一片蒟蒻凍片再放一顆紅豆草莓，拉起四個角包緊，倒放在盤中即可。

INGREDIENTS: 8-10 people
20g konjac powder, 400cc water, 200g sticky rice flour, 200cc water, 10 strawberries, 100g aduki bean paste, 3T cornstarch

METHODS:

1. Combine sticky rice flour with 200cc of water well, then pour into a shallow steaming plate. Steam for about 10 minutes and remove to set.
2. Combine konjac powder with 400cc of water to make into konjac liquid. Pour into a shallow plate until set to make into a thin konjac slice. Let cool and cut into 10 equal squares.
3. Divide the bean paste into 10 equal portions. Wrap one strawberry up in one portion of bean paste. Roll each one into a round ball.
4. Sprinkle cornstarch evenly over the sticky rice sheet, divide the sheet into 10 equal slices. Place one slice of konjac slice on each sticky rice slice first, then top with a stuffed bean ball. Pull up four corners and tie tightly. Transfer to serving plate and serve.

1. 富含大量維生素C的草莓一直是深受大家喜愛的水果，現在有了蒟蒻加草莓的新吃法，當然要試一試！
2. 大福多為手工製作，因此市售的單價偏高，不妨自己製作，只要花一點點材料費，就可以享受到高價的美味，更是招待朋友不可或缺的美味點心。

1. Strawberry contains abundant vitamin C and is everyone' s favorite fruit. Now this new dish with konjac and strawberry, you definitely have to try it out!
2. This homemade Tafu dish is quite expensive in the market. It is better that you try it yourself at home. With just a small expense for ingredients, you can enjoy high-priced, delicious food. It is a wonderfully delicious dessert for entertaining friends.

Banana in Konjac Rolls

Banana in 蒟蒻香蕉卷 Konjac Rolls

材料：（3~5人份）
蒟蒻凍粉20克、水500c.c.、香蕉3條

做法：
1.蒟蒻凍粉加水調勻煮沸後倒入淺盤中，待涼後切成3份，製作成蒟蒻薄片。
2.香蕉去皮用蒟蒻薄片包起來，切成適口的大小即可。

INGREDIENTS: 3-5 people
20g konjac powder, 500cc water,
3 bananas

METHODS:
1. Combine konjac powder with water and stir until evenly mixed. Cook in pan until boiling and pour over in a shallow plate until set. Then cut into 3 equal thin slices.
2. Remove skin from bananas and wrap each up with one konjac slice well, then cut into serving size. Serve.

1.做法簡單且好吃，媽媽們可以多做些給孩子們吃，只要花點心思，簡單的食物也可以變化出各式料理。
2.可利用方形淺盤製作蒟蒻薄片，取出後稍微切割即可使用。
3.香蕉要選比較直的，底部要切平才好置盤。如果想要多點味道，可以淋上果醬或鮮奶油。
4.香蕉是台灣的特產，非常容易購得。吃香蕉可以潤腸通便、降血壓；香蕉還含有豐富的鉀，根據研究報告指出，可以降低中風及高血壓的發生率。

1. This dish is simple and delicious to prepare. Mothers can prepare more for their kids. With only a fewe more changes, simple foods can become many kinds of delicious cuisine.
2. Use a square shallow plate for the konjac liquid, so that when it is set, you just simply cut it into slices as needed.
3. Selecting straighter bananas for this dish, and cut the bottom flat so that it is easy to arrange on the plate. If you prefer a little more flavor, drizzle with jam or whipping cream.
4. Bananas are Taiwan's specialty product. It is very easy to obtain. Bananas helps moisten the intestines and cure constipation, as well as lower blood pressure. Banana also contains abundant potassium. According to research, it reduces the incidence of strokes and high blood pressure.

Baked Konjac Snack Bars

Baked Konjac 蒟蒻點心條 Snack Bars

材料：（5~6人份）
熟蒟蒻塊600克

調味料：
薄鹽醬油1/2杯、水2杯、冰糖粉2大匙、八角10克、白胡椒粉1/2小匙

做法：
1.將去鹼的熟蒟蒻塊切成適當的條狀。
2.將蒟蒻條和所有調味料放入鍋中煮沸熄火，浸泡12小時使其充分入味。
3.將蒟蒻條連同汁液以小火慢煮至湯汁收乾。
4.取出蒟蒻條平鋪在錫箔紙上，放入烤箱中以120℃（248℉）全火烤20分鐘即可。

INGREDIENTS: 5-6 people
600g konjac strips

SESAONINGS:
1/2C light soy sauce, 2C water, 2T rock sugar powder, 10g star anise, 1/2t white pepper

METHODS:
1.Cut the cooked konjac, from which the alkali has already been removed, into appropriate size strips.
2.Heat konjac strips and all the seasonings in pan until boiling, remove from heat and soak for 12 hours until the flavor is completely absorbed.
3.Cook konjac in pan along with the seasonings over low heat until the seasonings are dry.
4.Remove konjac strips onto aluminum foil in baking sheet, bake in oven at 120℃（248℉） on both heat elements for 20 minutes. Remove and serve.

1.如果你覺得市售的蒟蒻零食調味過重、口感過硬的話，那麼就自己動手做吧！既經濟實惠又能兼顧健康。如果家裡沒有烤箱，將蒟蒻條放在室溫下風乾也可以。提醒你，吃的時候一定要細細咀嚼哦！
2.加入八角及胡椒粉可使蒟蒻條味道更香、吃起來更夠味。

1. If you think the konjac snack sold in markets is too heavily seasoned and the texture is too tough, then make it at home yourself. It is both economical and healthy. If an oven is not available, blow drying at room temperature is OK too! Remember to chew slowly to taste the flavor fully.
2. Konjac bars with star anise and pepper added are delicious and satisfying.

蒟蒻纖瘦健康吃
Healthy Slimming with Konjac

Simple Side Dishes 簡易小菜

三杯蒟蒻透抽Three-Cup Konjac Squid
薑絲炒脆腸Stir-Fried Konjac Intestine with Shredded Ginger
涼拌蒟蒻杏鮑菇Cold Konjac with Japanese Oyster Mushrooms
銀芽海參片Konjac Sea Cucumber with Bean Sprouts
梅漬彩椒蒟蒻片Konjac with Plum and Bell Peppers
蕃茄梅菇蒟蒻Plum Konjac with Cherry Tomato and Seaweed
泡菜凍Pickled Vegetable Konjac
花生素凍Peanut Konjac
香酥蒟蒻Crispy Konjac
素香藕片Vegetarian Konjac Lotus Root with Broccoli
繽紛蒟蒻絲Colored Shredded Konjac with Mushrooms and Bell Peppers
牛蒡素蜇絲Vegetarian Konjac Jellyfish with Burdock
什錦蒟蒻炒Well-Mixed Konjac Stir-Fry
燒烤蒟蒻串Barbecue Konjac Shiskaba
南瓜豆腐蒟蒻凍Pumpkin with Tofu Konjac
彩頭水晶凍Radish and Seaweed Konjac Jelly
莧菜豆腐腦兒Konjac Tofu with Amaranth
薯泥鮭魚片Potato Tower with Carrot Konjac in Pumpkin Gravy
紅麴蒟蒻Konjac with Red Wine Lees
蒟蒻魷魚排Fried Konjac Squid Steak with White Sesame

Three-Cup Konjac Squid

三杯蒟蒻透抽

材料：（5~6人份）
蒟蒻透抽300克、老薑片30克、九層塔適量、紅甜椒、黃甜椒各100克、紅辣椒粒適量

調味料：
醬油4大匙、胡麻油 1大匙、糖1大匙、水120.c.c

做法：
1.進口胡蘿蔔洗淨擦乾表面水分，取一塊生蒟蒻包住胡蘿蔔，接縫處要捏緊，放入水中加熱定型。煮熟後取出胡蘿蔔，將蒟蒻切成一圈圈，再依基本做法完成去鹼後即可調理。
2.九層塔洗淨瀝乾水分，紅、黃甜椒洗淨切片備用。
3.起油鍋，爆香老薑片，加入醬油、糖、水、蒟蒻透抽及辣椒粒，翻炒之後蓋上鍋蓋，以小火燜至湯汁收乾。
4.起鍋前，放入紅、黃甜椒、九層塔即可。

INGREDIENTS:

5-6 people
300g konjac squid, 30g old ginger slices, basil leaves as needed, 100g each red bell pepper and yellow bell pepper, diced red chili peppers as needed

SEASONINGS:

4T soy sauce, 1T black sesame oil, 1T sugar, 120cc water

METHODS:

1. Cut the cooked konjac, from which the alkali has already been removed, into appropriate size strips.
2. Rinse basil leaves well and drain well. Rinse red and yellow bell peppers well, then slice for later use.
3. Heat wok and stir-fry old ginger slices until fragrant. Add soy sauce, sugar, water and konjac squid along with chili pepper dices. Stir rapidly until well-mixed, cover and simmer over low until the liquid is absorbed.
4. Add red, yellow peppers and basil leaves to mix before removing from heat. Serve.

TIPS

1.這是非常下飯的一道料理，就算涼了也很好吃！
2.蒟蒻較不易咀嚼，細嚼慢嚥才能有助消化且享受好滋味。

1. This is a dish that stimulates the appetite and is still delicious even served cold.
2. Konjac is not easy to chew. Chewing slowly until fine helps digestion and enables you to enjoy the delicious taste.

Stir-Fried Konjac Intestine with Shredded Ginger

薑絲炒脆腸

材料：（6~8人份）
蒟蒻脆腸250克、酸菜100克、嫩薑絲60克、紅辣椒少許

調味料：
米醋3大匙、鹽1/4小匙、糖1/2小匙、油1小匙、水適量

做法：
1.酸菜切絲（如果過鹹須浸泡在水中去除鹹味）備用，嫩薑切絲浸在加了少許白醋的水中，保持脆度並去辛辣味。
2.起油鍋，放入酸菜絲、薑絲、脆腸、辣椒同炒，再加入其他調味料續炒即可。

INGREDIENTS:

6-8 people
250g konjac intestines, 100g pickled mustard, 60g shredded tender ginger, red chili pepper as needed

SEASONINGS:

3T rice vinegar, 1/4t salt, 1/2t sugar, 1t cooking oil, water as needed

METHODS:

1. Shred pickled mustard (soak in water first to remove saltiness if too salty) and set aside. Shred tender ginger and soak in vinegar water to maintain crunchiness and remove its spiciness.
2. Heat oil in wok, stir-fry shredded pickled mustard, shredded ginger, konjac intestine and chili pepper for a minute, then season with the remaining seasonings. Continue stirring until done. Serve.

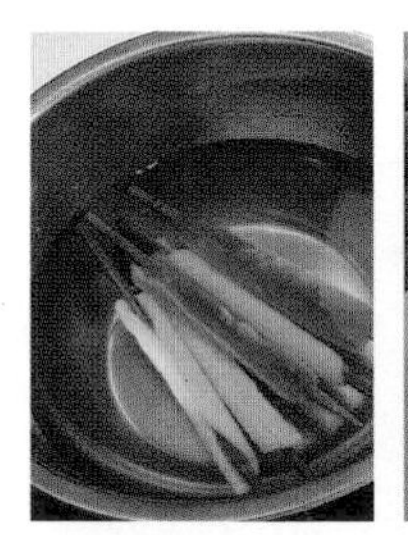

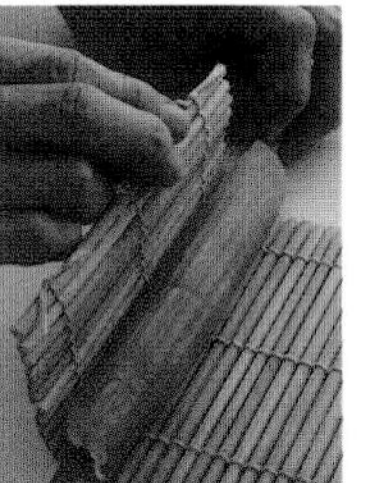

蒟蒻脆腸做法請見P.15

TIPS

1.這道菜因加入重口味的酸菜和辣椒，很適合做為夏天的開胃菜，可促進食慾，喜歡吃辣的人，更不可放過，美味絕對令你難忘。
2.不敢吃辣的人，可將辣椒籽去除再加入，取辣椒鮮豔的色澤入菜即可。此外，吃素者不宜吃這道菜。

1. This dish with heavy flavor ingredients like pickled mustard and chili pepper is an excellent appetizer for the summer. It improves the appetite. Those who enjoy spicy food should not miss it. It can be served as a side dish with wine or desserts. You will never forget its delicious flavor.
2. If spicy food is not desired, discard the chili pepper seeds and use bright red chili pepper in dish.Furthermore, this dish is not suitable for vegetarians.

Cold Konjac with Japanese Oyster Mushrooms

材料： （5~6人份）
枸杞蒟蒻塊150克、小黃瓜1條、杏鮑菇100克、辣椒絲少許、香菜少許

調味料：
鎮江醋1大匙、薄鹽醬油1小匙、糖1/2小匙、冷開水1/2杯、鹽1/4小匙、香油1/4小匙

做法：
1. 枸杞10克泡軟，加水900 c.c.打成汁，再依「蒟蒻塊自己做」的做法1、2做成枸杞蒟蒻塊。
2. 蒟蒻塊、小黃瓜都切成長約3公分的條狀，用鹽醃約15分鐘，再用冷開水洗去鹽分備用。
3. 杏鮑菇切成條狀，放入沸水中煮約1分鐘，取出後瀝去水分並放涼。
4. 將醬油、醋、糖、冷開水調勻後拌入蒟蒻、小黃瓜、杏鮑菇及辣椒絲、香菜等浸泡30分鐘使其入味，盛盤前再滴些許香油即可。

INGREDIENTS:

5-6 people
150g lycium berry konjac pieces, 1 Chinese cucumber, 100g Japanese oyster mushrooms, shredded chili pepper and cilantro as needed

SEASONINGS:

1T Chen-ching vinegar, 1t light-salt soy sauce, 1/2t sugar, 1/2C cold water, 1/4t salt, 1/4t light sesame oil

METHODS:

1. Soak 10g of lycium berries until soft, mix with 900cc of water and blend to form juice. Then follow the methods in "Konjac Chunks Do It Yourself" on p.14 to prepare lycium berry konjac chunks.
2. Cut konjac pieces and Chinese cucumber into 3cm long strips, then marinate in salt for about 15 minutes and rinse out with cold water.
3. Cut Japanese oyster mushrooms into strips and cook in boiling water for about 1 minute. Remove and drain well, then let cool.
4. Combine soy sauce, vinegar, sugar and cold water well, then add konjac strips, Chinese cucumber strips, Japanese oyster mushrooms along with shredded chili peppers and cilantro. Mix well and soak for 30 minutes until the flavor is well absorbed. Transfer to serving plate and drizzle with a little light sesame oil. Serve.

1. 杏鮑菇吃起來有點杏仁的香味，口感像鮑魚，可稍加調味料清炒即可。
2. 不喜歡吃辣的人，可以不要放辣椒絲。

1. Japanese oyster mushrooms have the flavor of almond with the texture of abalone. It may simply be stir-fried with a little seasoning added.
2. Shredded chili pepper can be skipped if not desired.

Konjac Sea Cucumber with Bean Sprouts

銀芽海參片

材料：（3~5人份）
蒟蒻海參片200克、綠豆芽200克、嫩薑絲少許

調味料：
A.糖1小匙、醬油1小匙、烏醋2大匙、水100c.c.
B.油1/2小匙

做法：
1.乾木耳10克泡軟，再依p.14「蒟蒻塊自己做」做法1、2做成蒟蒻海參片。
2.將綠豆芽摘去頭尾，洗淨瀝乾水分備用。
3.用1/2小匙油將嫩薑絲爆香，放入調味料、已經完全去鹼的蒟蒻海參片，轉小火燜煮2分鐘，再轉成大火並放入綠豆芽快炒1分鐘即可。

INGREDIENTS:
3-5 people
200g konjac sea cucumber slices, 200g mung bean sprouts, shredded young ginger as needed

SEASONINGS:
A: 1t sugar, 1t soy sauce, 2T black vinegar, 100cc water
B: 1/2t cooking oil

METHODS:
1. Soak 10g of dried woodear fungus until soft, then follow the methods in "Konjac Chunks Do It Yourself" on p.14 to prepare sea cucumber konjac slices.
2. Remove both ends from mung bean sprouts and discard, rinse well and drain.
3. Heat 1/2t of cooking oil in wok and stir-fry shredded young ginger until the flavor is released, season with seasonings along with the konjac cucumber slices, which alkali has been removed. Reduce heat to low and simmer for 2 minutes, then increase heat to high and saute mung bean sprouts rapidly for 1 minute. Remove and serve.

1.小火燜煮蒟蒻海參可使其更加入味，如果已經事先做好浸漬的準備工作，這個步驟就可以省略。
2.綠豆芽摘掉頭尾部分即銀芽，綠豆芽是高纖食物，可以促進腸子蠕動，減少體內脂肪堆積。

1. Simmering konjac sea cucumber over low heat enables the flavor to be absorbed better. If marinating is done for a long enough time, this step can be skipped.
2. Remove both ends of mung bean sprouts to make this dish. Mung bean sprouts are a high fiber food that helps the intestines move and reduces the storage of fat in the body.

Konjac with Plum and Bell Peppers

梅漬彩椒蒟蒻片

材料：（5~6人份）
梅肉20克、各色甜椒200克、蒟蒻片100克、冷開水100 c.c.

做法：
1.將蒟蒻片和梅肉泡在冷開水中約30分鐘。
2.甜椒洗淨，去籽切成薄片，放入沸水中燙10~15秒後撈起，立即泡入冰開水中3分鐘，瀝乾水分備用。
3.將甜椒片均勻浸漬在做法1.的梅汁中，放入冰箱冷藏，待入味且冰透後即可取出食用。

INGREDIENTS:

5-6 people
20g plum flesh, 200g colored bell peppers, 100g konjac slices, 100cc cold water

METHODS:

1. Soak konjac and plum in cold water for about 30 minutes.
2. Rinse bell peppers, discard the seeds and cut into thin slices. Blanch in boiling water for 10 to 15 seconds and remove to soak in ice water for 3 minutes. Then drain well.
3. Soak bell pepper slices in method(1) plum juice evenly, then chill in the refrigerator until flavor is absorbed and icy cold. Remove and serve.

1.做法簡單、材料準備方便的涼拌菜既開胃又爽口，夏天缺乏食慾時，不妨多做些。
2.梅子是最養生的零嘴，但選購時，要注意避免含有色素和添加物的產品。

1. This delicious appetizer is simple and easy to prepare, the ingredients are convenient to obtain. When your appetite is poor in the summer, this is a perfect dish to make.
2. Plum is a nourishing snack. Select plums without food coloring and other flavorings.

Plum Konjac with Cherry Tomato and Seaweed

蕃茄梅菇蒟蒻

材料：（5~6人份）
蒟蒻片150克、小蕃茄150克、海帶芽5克、鴻喜菇100克、梅肉10克

調味料：
油1/2小匙、水1/2杯、薑末1/4小匙

做法：
1. 蒟蒻片加水、梅肉、薑末一起浸泡20分鐘。
2. 海帶芽用冷水泡開後瀝乾水分，鴻喜菇洗淨切去根部一朵一朵分開，備用。
3. 小蕃茄洗淨去蒂對切，用油煎軟。放入做法1.及海帶芽、鴻喜菇煮3分鐘即可。

INGREDIENTS:

5-6 people
150g konjac slices, 150g cherry tomatoes, 5g dried wakame seaseeds, 100g maitake mushrooms, 10g plum flesh

SEASONINGS:

1/2t cooking oil, 1/2C water, 1/4t minced ginger

METHODS:

1. Combine konjac slices with water, plum flesh and minoed ginger together, then soak for 20 minutes.
2. Soak wakame seaweed in cold water until swollen and drain well. Rinse mushrooms well, discard the stems and separate them apart.
3. Rinse cherry tomatoes, discard stems and cut each in half. Fry with oil until soft and then add method (1) along with wakame seaweed and maitake mushrooms. Cook for 3 minutes and remove. Serve.

TIPS

1. 低油脂的料理一向都很簡單操作，符合忙碌上班族的需求。
2. 平日做菜時，不妨多選用海帶芽來搭配其他食材入菜，多吃海帶芽可防止骨質流失，預防骨質疏鬆症，多存一些骨本。

1. Low fat cuisine is always very easy to handle and is perfect for the needs of busy career women.
2. Use wakame seaweed frequently as an accompaniment in dishes because it prevents loss of bone density and osteoporosis.

Pickled Vegetable Konjac

泡菜凍

材料：（6~8人份）
蒟蒻凍粉40克、水1000c.c.、泡菜200克

做法：
1.蒟蒻凍粉加水攪拌均勻。
2.泡菜切碎一起放入蒟蒻凍液中，以中火煮沸。
3.將做法2.倒入耐熱模型中，在室溫下放涼後移入冰箱，約50分鐘取出切塊食用。

INGREDIENTS:

6-8 people
40g konjac powder, 1000cc water, 200g pickled vegetables

METHODS:

1. Combine konjac powder with water well, stir until evenly mixed.
2. Chop pickles finely and mix well with konjac water from method (1). Bring to boil over medium heat.
3. Pour method (2) into heat resistant molds. Let sit at room temperature until cool, then remove to refrigerator and chill for about 50 minutes until set. Remove and cut into serving sizes. Serve.

1.泡菜低熱量又開胃，用它來搭配蒟蒻做成夏日小菜，方便又可口。這道小菜口味清淡，喜歡重口味的人可以另外準備醬油、辣椒、香油的沾醬。
2.台灣蔬菜種類豐富，整年都可以吃到新鮮的蔬菜，不妨選擇適當的蔬菜自己製作泡菜，自己動手做既衛生，還可一年四季都吃得到。

1. Pickles are low calories and stimulate the appetite. Combining them with konjac to make summer side dishes is convenient and delicious. This dish is very light. For those who prefer heavy flavor, the dipping sauce can be served on the side with soy sauce, chili pepper or light sesame oil.
2. There are many kinds of vegetables in Taiwan, and fresh vegetables may be served all year round. Selecting seasonal vegetables to make into pickles you prepare yourself will be clean, healthy dishes that are ready to be served any time of the year.

Peanut Konjac

花生素凍

材料：（8~10人份）
蒟蒻凍粉40克、水1000c.c.、水煮花生仁200克、乾海帶芽10克、嫩薑1小塊

調味料：
鹽1/2小匙、香菇粉1/2小匙、淡醬油1大匙、香油少許

做法：
1.乾海帶芽泡軟，加水用果汁機打成汁。
2.海帶汁內加入蒟蒻凍粉、香菇粉、鹽調味，煮至約70°C（158°F）時加入花生仁，續煮至微微沸騰後熄火。
3.將煮好的蒟蒻凍倒在耐熱容器中放涼，切成塊狀。
4.嫩薑磨泥和淡醬油調成沾醬，食用時邊沾醬汁即可。

INGREDIENTS:

8-10 people
40g konjac powder, 1000cc water, 200g water-boiled shelled peanuts, 10g dried wakame seaweed, 1 small piece tender ginger

SEASONINGS:

1/2t salt, 1/2t shiitake mushroom powder, 1T light soy sauce, light sesame oil as needed

METHODS:

1. Soak dried wakame seaweed in water until soft, then blend in blender with water added to make wakame water.
2. Add konjac powder, shiitake mushroom powder and salt to wakame water. Cook until the temperature reaches 70°C (158°F), add peanuts, then continue cooking until lightly boiling and remove from heat.
3. Pour the konjac liquid into heat resistant molds to set, then cut into pieces.
4. Mash young ginger well. Mix well with light soy sauce to make into dipping sauce and serve on the side.

TIPS

花生含卵磷脂和腦磷脂，可以緩和腦部衰退，對於兒童及成長期青少年，可幫助提昇記憶力。

Peanuts contains lecithin and phosphatidyl ethanol. It can slow brain deterioration, and helps enhane the memory of children and teenagers.

Crispy Konjac

香酥蒟蒻

材料： （5~6人份）
滷味蒟蒻條200克、馬鈴薯200克、低筋麵粉100克.、冷開水100 c.c.、油240 c.c.

做法：
1.蒟蒻塊切厚條狀，在滷汁中浸泡12小時後取出，即成滷味蒟蒻條。
2.馬鈴薯去皮切細絲泡入水中，大約換水2~3次後撈出，瀝乾水分備用。
3.取一大碗，將麵粉、冷開水調成漿，再放入馬鈴薯絲拌勻。
4.取適量馬鈴薯絲包住蒟蒻條，放入八分熱的油鍋炸酥即可。

INGREDIENTS:
5-6 people
200g stewed konjac Julienne, 200g potatoes, 100g cake flour, 100cc cold water, 240cc cooking oil

METHODS:
1. To prepare stewed konjac strips - Cut konjac chunks into thick strips, soak in stewing liquid for 12 hours and remove.
2. Peel potatoes and shred finely, then soak in water. Change the water two or three times, then remove from water and drain well.
3. Combine flour and cold water in a large mixing bowl to make flour batter, then add shredded potatoes and mix well.
4. Take a suitable amount of shredded potatoes in hand and wrap up konjac julienne in center. Deep-fry in 80% steaming oil until crispy and remove. Serve.

1.多做一些滷味蒟蒻，就可以用一部分來嘗試這種不一樣的吃法。蒟蒻容易入味、口感酥脆……小心停不下來喲！
2.油炸的烹調方式容易攝取過多的脂肪，偶爾解饞就好，不可以常吃。

1. Prepare more stewed konjac and store away, then you can use part of it to make other kinds of dishes. Once that tasty konjac with its crispy texture hits your mouth, be careful, because you might not be able to stop!
2. It is easy to take in too much fat when the food is deep-fried, so prepare once in a while.

Vegetarian Konjac Lotus Root with Broccoli

素香藕片

材料：（3~5人份）
蒟蒻藕片150克、綠花椰菜100克

調味料：
白味噌1大匙、原味優格2大匙、冰糖粉1/2小匙、白芝麻1/4小匙

做法：
1.取一大碗，將所有調味料調勻，拌入蒟蒻藕片浸泡約50分鐘。
2.綠花椰菜洗淨，放入沸水中燙熟後取出漂冷開水，瀝乾水分後一起拌入調味醬中即可。

INGREDIENTS:
3-5 people
150g konjac lotus root slices, 100g broccoli

SEASONINGS:
1T light miso paste, 2T plain yogurt, 1/2t rock sugar powder, 1/4t white sesame seeds

METHODS:
1. Combine all the seasonings in mixing bowl well, then soak konjac lotus root slices in seasonings for about 50 minutes until flavor is well absorbed.
2. Rinse broccoli well, blanch in boiling water until done, then remove and soak in cold water for a minute. Drain well and add to the marinade. Serve.

TIPS

再次提醒你：做好的蒟蒻藕片要事先浸泡在調過味的汁液中才能讓蒟蒻變得好吃，所以，這道菜可以多做一些放在冰箱裡，搭配各類適合的蔬果就能隨時上桌。

To remind you once again, the konjac lotus root slices have to be soak in the well-mixed seasonings first so that they will be tasty. This dish can be prepared in large amounts and stored in the refrigerator. It is perfect for all kinds of vegetables and can be served any time.

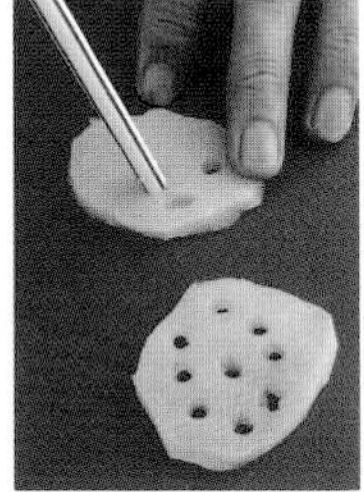

蒟蒻藕片做法參見P.16

Colored Shredded Konjac with Mushrooms and Bell Peppers

Colored Shredded Konjac with

繽紛蒟蒻絲

Mushrooms and Bell Peppers

材料：（3~5人份）
芝麻蒟蒻絲50克、胡蘿蔔蒟蒻絲50克、黃椒蒟蒻絲50克、金針菇100克、紅甜椒100克

調味料：
油1/2小匙、鹽1/4小匙、水2大匙

做法：
1.將3大匙無糖黑芝麻粉加入900c.c.的水中，再依p.14「蒟蒻塊自己做」的做法做成芝麻蒟蒻塊，切成細絲，即芝麻蒟蒻絲。
2.金針菇洗淨一絲絲分開，紅甜椒洗淨去籽切絲。
3.起油鍋炒金針菇、紅甜椒，放入各色蒟蒻絲、鹽、水續炒約3分鐘即可。

INGREDIENTS: 3-5 people
50g shredded sesame konjac, 50g shredded carrot konjac, 50g shredded yellow bell konjac, 100g velvet foot mushrooms, 100g red bell peppers

SEASONINGS:
1/2t cooking oil, 1/4t salt, 2T water

METHODS:
1.Combine 3T of unsweetened black sesame powder with 900cc of water, then follow the methods in "Konjac Chunks Do It Yourself" on p.14 to prepare sesame konjac chunks. Then shred finely to make shredded sesame konjac.
2.Rinse velvet foot mushrooms well and separate them one by one. Rinse red bell pepper, discard seeds and shred.
3.Heat oil in wok and stir-fry velvet foot mushrooms and red bell peppers along with all the shredded colored konjac. Season with salt and add water. Stir-fry for about 3 minutes and remove. Serve.

這道菜的顏色豐富極能增進食慾，你可以在五種顏色當中最多使用三種蒟蒻絲就好，其他可用蔬菜或菇類來搭配，才能攝取到完整的養分。

This colorful dish increases the appetite. Use konjac for three of the five different colors in this dish, then use vegetables or mushrooms for the remainder, for more complete nutrition.

Vegetarian Konjac Jellyfish with Burdock

Vegetarian Konjac

牛蒡素蜇絲

Jellyfish with Burdock

材料：（5~6人份）
蒟蒻海蜇絲200克、牛蒡1支、白芝麻1大匙

調味料：
醬油3大匙、二砂糖2小匙、油2小匙、水100c.c.

醋水：
白醋1大匙、水1000C.C.

做法：

1.生蒟蒻放在室溫下至不沾手時，每次取一小塊用手掌壓成薄片，放入沸水中煮熟，去鹼後切絲即成蒟蒻海蜇絲。
2.將1小匙油倒入鍋中，放入醬油、糖，和水一起煮開，把蒟蒻絲浸泡在醬汁中30分鐘入味。
3.牛蒡去皮切細絲，將切好的牛蒡絲泡入醋水中避免牛蒡變黑。
4.起油鍋，炒香牛蒡絲，放入醬汁及蒟蒻絲同炒至水分收乾。
5.起鍋前撒下白芝麻即可。

INGREDIENTS: 5-6 people
200g konjac jellyfish, 1 burdock stick, 1T white sesame seeds

SEASONINGS:
3T soy sauce, 2t no.2 granulated sugar, 2t cooking oil, 100cc water

VINEGAR WATER:
1T white vinegar, 1000cc water

METHODS:

1. Let uncooked konjac sit at room temperature until no longer sticky. Remove a little at a time and press flat between palms, then cook in boiling water until done. Remove to soak in water to remove the alkali, then shred finely to make shredded jellyfish konjac.
2. Cook 1t of cooking oil in pan to a boil with soy sauce, sugar and water added, then soak shredded konjac for 30 minutes until flavor is absorbed.
3. Peel burdock and then shred finely, soak in vinegar water to prevent it from darkening.
4. Heat oil in wok and stir-fry shredded burdock until fragrant. Add marinade along with shredded konjac, stir rapidly until the liquid is totally absorbed.
5. Sprinkle with white sesame seeds before removing from heat.

1.牛蒡絲如果切得細一點會更好吃！
2.夏天可以多做一些放冰箱冷藏，就可以隨時吃到可口的小菜囉！
3.牛蒡是根菜類植物，含大量植物纖維，可發揮改善便祕的功效。

1.The more finely the burdock is shredded, the better it will taste.
2.Prepare as much as you can and store in the refrigerator, so that this delicious dish can be served any time.
3.Burdock is a kind of root plant and contains large quantity of plant fiber. Serving often can help to with constipation.

Well-Mixed Konjac Stir-Fry

Well-Mixed

什錦蒟蒻炒

Konjac Stir-Fry

材料：（3~5人份）
蒟蒻脆腸100克、竹筍50克、西洋芹70克、乾香菇15克、金針10克

調味料：
鹽1/4小匙、水1大匙、油1/2小匙

做法：
1.竹筍剝去外殼切成條狀，西洋芹去除較粗纖維後先切段，再將每段切成2片，一起入沸水中燙熟取出。
2.乾香菇與金針洗淨泡軟，香菇去蒂切條狀。
3.將竹筍、香菇、西洋芹、金針各一份塞入蒟蒻脆腸中。
4.起油鍋，放入脆腸卷、鹽及水快炒2分鐘即可。

INGREDIENTS: 3-5 people
100g konjac intestines, 50g bamboo shoots, 70g western celery stalks, 15g dried shiitake mushrooms, 10g dried day lily flowers,

SEASONINGS:
1/4t salt, 1T water, 1/2t cooking oil

METHODS:
1. Remove outer shells from bamboo shoots and cut into strips. Tear off the coarse skin from celery and cut into sections, then cut each section into 2 slices. Cook together in boiling water until done and remove.
2. Rinse dried shiitake mushrooms and dried day lily flowers until softened, then remove stems from mushrooms.
3. Stuff bamboo shoots, mushrooms, celery and day lily flowers inside the konjac intestines.
4. Heat oil in wok, add salt, and rapidly stir-fry stuffed konjac intestines for 2 minutes until done. Serve.

這道菜也可以加點酸菜絲、嫩薑絲一起煮成湯，夏日炎炎裡，是一道可以開胃的好湯。

Shredded pickled mustard or shredded young ginger can be added to this dish, which can be made into a soup. On hot summer days, this is an excellent soup for stimulating the appetite.

Barbecue Konjac Shiskaba

Barbecue Konjac Shiskaba

燒烤蒟蒻串

材料：（3~4人份）
蒟蒻片200克、鮮香菇100克、各色甜椒100克、五香豆乾180克

調味料：
醬油1大匙、糖1大匙、沙茶醬3大匙、冷開水120c.c.、白芝麻1/2小匙

做法：
1. 將蒟蒻片的一面用交叉方式刻出花紋，豆乾洗淨切塊，用沸水燙熟備用。
2. 蒟蒻片、豆乾浸泡在所有調味料中60分鐘，取出瀝乾。
3. 香菇、甜椒洗淨後切成易入口的塊狀，和豆乾依序包在蒟蒻片中，再以竹籤串起來。
4. 取一平底鍋，鍋中不放油乾煎蒟蒻串，不時淋上調味料及翻面，大約煎5分鐘即可。

INGREDIENTS: 3-4 people
200g konjac slices, 100g fresh shiitake mushrooms, 100g all kinds of colored bell peppers, 180g five-spiced pressed tofu

SEASONINGS:
1T soy sauce, 1T sugar, 3T barbecue sauce, 120cc cold water, 1/2t white sesame seeds

METHODS:
1. Score Xs on one side of konjac slices and set aside. Rinse pressed tofu and cut into pieces, then blanch in boiling water until done.
2. Soak konjac slices and pressed tofu in well-combined seasonings for 60 minutes, then remove and drain.
3. Rinse shiitake mushrooms and bell peppers well, then cut into serving sizes. Wrap up in konjac slices along with pressed tofu, then thread onto bamboo skewers.
4. Heat a frying pan until smoking, then dry-fry konjac shish kebabs. Drizzle with the seasonings and turn constantly. Fry for about 5 minutes and serve.

1.此道菜用乾煎的方式也一樣好吃，而且比較符合健康概念。事先浸泡蒟蒻片則可讓它更入味！
2.這道菜什麼顏色都有，做法簡單不需耗費太多時間，可當做朋友來訪時的下酒菜或點心。

1. Though this dish is called "barbecue," in fact dry-frying will produce the same delicious flavor, and be more healthy. Soaking konjac slices beforehand will make the flavor be more deeply absorbed.
2. This dish contains a variety of colors and is easy to prepare without wasting too much time. When friends come over to visit, it is a perfect dish with wine or for dessert.

Pumpkin with Tofu Konjac

Pumpkin with Tofu Konjac

南瓜豆腐蒟蒻凍

材料：（8~10人份）
蒟蒻凍粉40克、黑糖80克、南瓜300克、豆腐1/2塊、水1000c.c.

調味醬汁：
素蠔油2大匙、冷開水1大匙、薑泥少許

做法：
1.將南瓜蒸熟，用湯匙取出南瓜肉；豆腐燙過沸水後切塊備用。
2.蒟蒻凍粉加入冷水中調勻，以小火煮至約60°C（140°F）時加入黑糖續煮，80°C（176°F）時再放入南瓜及豆腐同煮（須輕輕攪動避免沾鍋），微微沸騰時即可熄火。
3.將蒟蒻凍液倒入方形盤中，待涼切塊淋上調好的醬汁即可。

INGREDIENTS: 8-10 people
40g konjac powder, 80g dark brown sugar, 300g pumpkin, 1/2 tofu, 1000cc water

SAUCE FOR DRIZZLING:
2T vegetarian oyster sauce, 1T cold water, mashed ginger as needed

METHODS:
1. Steam pumpkin until done and scoop out the flesh with a spoon. Blanch tofu with boiling water and cut into pieces.
2. Combine konjac powder with cold water well. Cook over low until the temperature reaches 60°C (140°F), then add dark brown sugar to taste. Continue cooking until it reaches 80°C (176°F), add pumpkin flesh and tofu (stir gently to prevent from sticking to the bottom of the pan). Cook until lightly boiling and remove from heat.
3. Pour the konjac liquid into a square mold to cool and set. Slice and drizzle with well-mixed sauce. Serve.

1.黑糖加薑汁本來就很對味，再配上南瓜、豆腐……，視覺味覺都得到滿足了！
2.也可以用地瓜取代南瓜，但記得別蒸得過爛，才能保持彈性。

1. Dark brown sugar is perfect with ginger juice. With pumpkin and tofu added, the senses of taste and the sight are both completely satisfied.
2. Pumpkin can be substituted with yam, but do not oversteam or it will be chewy.

Radish and Seaweed Konjac Jelly

Radish and Seaweed

彩頭水晶凍

Konjac Jelly

材料：（6~8人份）
蒟蒻凍粉40克、醃漬蘿蔔干100克、乾海帶芽10克、水1000c.c.

做法：
1.蘿蔔干稍微切碎，海帶芽泡軟切碎。
2.蒟蒻凍粉加水攪拌均勻，加入蘿蔔干及海帶芽，以中火煮至微微沸騰即熄火。
3.將做法2.倒入淺盤中，待涼後切塊食用。

INGREDIENTS: 6-8 people
40g konjac powder, 100g shredded pickled daikon radish, 10g dried wakame seaweed, 1000cc water

METHODS:
1.Chop shredded radish lightly. Soak wakame seaweed in water until softened and chop finely.
2.Combine konjac powder in water and stir until evenly mixed. Add shredded pickled radish and wakame seaweed. Heat on medium until lightly boiling, then remove from heat.
3.Pour method (2) over on a shallow plate to cool and set, then cut into pieces and serve.

如果你是擅長醃漬醬菜的高手，那青木瓜、大頭菜、榨菜……等，都可以和蒟蒻凍粉一起搭配，做成各式的鹹凍。

If you are an expert in pickling vegetables, then all kinds of salty konjac jelly can be prepared by adding things like green papaya, kohlrabi or Szechwan mustard stem pickle to konjac powder.

Konjac Tofu with Amaranth

Konjac Tofu
莧菜豆腐腦兒
with Amaranth

材料：（3~4人份）
蒟蒻凍粉10克、無糖豆漿500c.c.、莧菜200克、水1/2杯、太白粉水2大匙

調味：
鹽1/4小匙、白芝麻油適量

做法：
1. 蒟蒻凍粉加豆漿調勻，以小火煮沸熄火，在室溫下放涼。
2. 莧菜洗淨略切，加水打成汁，放入鍋中煮沸、調味、勾成芡汁後熄火。
3. 將蒟蒻凍片成薄片，放入鍋中與莧菜芡汁均勻混和一下，再滴些白芝麻油增添香氣，最後盛入深盤中即可。

INGREDIENTS: 3-4 people
10g konjac powder, 500cc unsweetened soybeans milk, 200g Chinese amaranth, 1/2C water, 2T cornstarch water

SEASONINGS:
1/4t salt, white sesame oil as needed

METHODS:
1. Combine konjac powder with soybean milk well, bring to boil over low heat and remove from heat, then let cool at room temperature.
2. Rinse Chinese amaranth well and cut into several sections. Blend in blender with some water added to make into liquid. Remove to pan and bring to a boil, then season and thicken with cornstarch water. Remove from heat.
3. Out konjac tofu into thin slices and add to pan to mix well with Chinese amaranth. Drizzlc with some white sesame oil to enhance the flavor and transfer to a deep serving plate.Serve.

1. 用蒟蒻凍入菜時，要注意不可持續加熱以免融化。這樣即使在冷冷的秋冬，也能享受一碗熱騰騰的蒟蒻料理。
2. 豆漿不加糖，營養不扣分，怕胖的人也可以放心喝，不必計較糖分的攝取。

1. When using konjac in dishes, just be careful not to heat continuously to prevent it from melting. Then even in the cold autumn or on winter days, you can still enjoy a bowl of steaming delicious konjac.
2. Unsweetened soybean milk is as nutritious as sweetened. Even when you are on a diet, you can still enjoy it and not worry about the intake of sugar.

Potato Tower with Carrot Konjac in Pumpkin Gravy

Potato Tower with Carrot Konjac

薯泥鮭魚片

in Pumpkin Gravy

材料：（2~3人份）
蒟蒻凍粉20克、胡蘿蔔汁400c.c.、胡蘿蔔30克、馬鈴薯150克、南瓜100克、熟豌豆仁30克

調味料：
鹽1/2小匙、黑胡椒粉1/4小匙、水1杯

做法：
1.將蒟蒻凍粉、胡蘿蔔汁、鹽攪拌均勻後煮沸熄火，倒入淺盤中放涼。
2.將胡蘿蔔、馬鈴薯、南瓜一同入蒸鍋裡蒸熟。
3.馬鈴薯、胡蘿蔔壓成泥，加入豌豆仁、鹽、黑胡椒粉調味做成塔狀。
4.南瓜泥加水煮成稠狀，加少許鹽調味。
5.胡蘿蔔凍用刀切成長條薄片狀，圍在做法3.外圍一圈，再淋上南瓜汁即可。

INGREDIENTS: 2-3 people
20g konjac powder, 400cc carrot juice, 30g carrot, 150g potatoes, 100g pumpkin, 30g cooked peas

SEASONINGS:
2t salt, 1/4t black pepper, 1C water

METHODS:
1. Combine konjac powder, carrot juice and salt together in pan. Bring to a boil and remove from heat. Pour over a shallow plate to cool.
2. Cook carrot, potatoes and pumpkin in steamer until done.
3. Mash potatoes and carrot, add peas along with salt and black pepper added. Shape into a tower.
4. Cook mashed pumpkin and water until thickened, season with salt to taste.
5. Cut carrot konjac jelly into long thin strip with a knife and surround the tower from method (3), then drizzle with thicken pumpkin soup. Serve.

這是一道充滿視覺享受的西式料理，做慣中式菜餚，偶爾換換口味，才能增加生活樂趣。但要注意的是，鹽分別用在蒟蒻凍、馬鈴薯泥和南瓜汁裡，成品的口味才能偏淡，可自己斟酌用量。

This is more like a western style cuisine with more visual enjoyment. If you are used to Chinese cuisine, a change to this flavor will add some zest to life. Only one thing that must be carefully watched is the salt. The dish is more or less a light dish with the only salt in the konjac jelly, mashed potato and pumpkin soup. If you prefer a heavier flavor, add more salt as desired.

Konjac with Red Wine Lees

Konjac with Red Wine Lees

紅麴蒟蒻

材料：（5~6人份）
蒟蒻片200克

調味料：
紅麴3大匙、黑糖2大匙、醬油膏1大匙、嫩薑末1/2小匙、油1/4小匙

做法：
1. 起油鍋，把紅麴等調味料和蒟蒻片一起用小火煮沸，熄火放涼。
2. 取一乾淨的保鮮容器，將紅麴蒟蒻放入開始浸漬，大約1小時後入味即可食用。

INGREDIENTS: 5-6 people
200g konjac slices

SEASONINGS:
3T red wine lees, 2T dark brown sugar, 1T thicken soy sauce, 1/2t minced tender ginger, 1/4t cooking oil

METHODS:
1. Heat oil in wok, bring all the seasonings along with konjac slices to boil in pan over low heat. Remove to cool.
2. In a large container, soak konjac slices in seasonings for about 1 hour until the flavor is well absorbed. Ready to serve.

紅麴是道地客家料理中經常會被用到的調味料，用它來醃漬蒟蒻，讓原本無味的蒟蒻片加分不少，味道很特別，你一定要試試。

Red wine lees are a typical condiment used in Hakka cuisine. Using it to marinate the konjac enhances the latter's plain flavor, giving it a special flavor.

Fried Konjac Squid Steak with White Sesame

Fried Konjac Squid Steak

蒟蒻魷魚排

with White Sesame

材料： （4~5人份）
魷魚排2份、白芝麻1/4小匙、油1/4小匙

醬汁：
醬油2大匙、糖1大匙、水1/4杯、嫩薑泥1/2小匙、辣椒粉1/4小匙、香油1/2小匙

做法：
1. 取一深碗，在深碗中將所有的醬汁材料調勻。
2. 將魷魚排浸泡在醬汁中30~50分鐘使其入味。
3. 取一平底鍋，鍋中放油，放入魷魚排，淋上約1大匙的醬汁以小火慢煎，待醬汁略收乾時翻面再淋一次醬汁慢煎。
4. 撒上白芝麻即可盛盤。

INGREDIENTS: 4-5 people
2 portions squid steaks, 1/4t white sesame seeds, 1/4t cooking oil

Seasonings for Preparing Sauce:
2T soy sauce, 1T sugar, 1/4C water, 1/2t mashed young ginger, 1/4t chili powder, 1/2t light sesame oil

METHODS:
1. Combine all the seasonings for the sauce in a mixing bowl, stir until evenly mixed.
2. Soak squid steaks in sauce for 30 to 50 minutes until flavor is well absorbed.
3. Heat oil in frying pan and fry squid steaks with 1T of sauce drizzled over. Fry over low heat until the sauce is almost absorbed, turn over and drizzle with sauce again. Continue frying slowly over low heat until done.
4. Sprinkle with white sesame seeds and remove to serving plate. Serve.

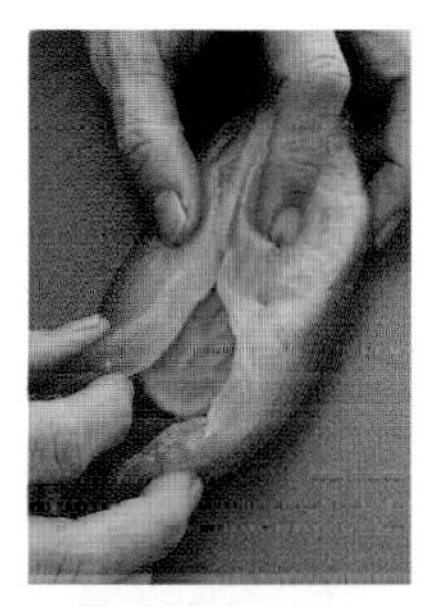

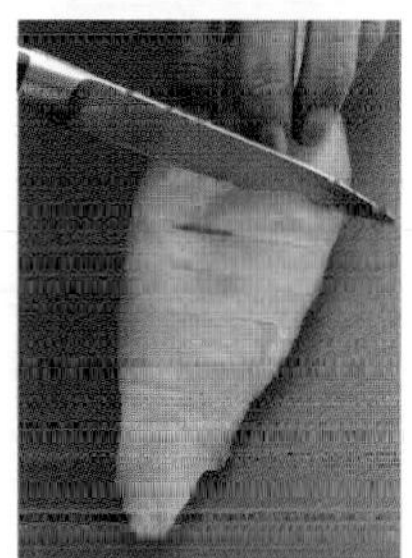

蒟蒻魷魚排做法
請見P.16

這是一道挺開胃的前菜。完全去鹼的蒟蒻成品，只要經過浸泡就能入味了，所以你可以調製各種喜愛的醬汁來做口味的變化。

This is an appetizer that really opens up your appetite. Soak konjac, after alkali has been soaked away, in sauce to absorb the flavor. For a change of pace, any flavor of sauce can be used for soaking the konjac.

Diet * Meal 瘦身大餐

蒟蒻蝦沙拉Konjac Shrimp Salad
鮮蔬咖哩蒟蒻Fresh Veggie Curry Konjac
栗燒蒟蒻干貝Konjac Scallop and Chestnuts
煎鑲甜椒Baked Stuffed Bell Peppers
南瓜蒟蒻扣Flip-Over Pumpkin Konjac
蝦仁鍋巴Konjac Shrimp over Burnt Rice
蒟蒻燴福球Konjac Tofu Balls with Bok Choy
山藥蒟蒻海棠煮Chinese Yam Konjac with Loofah and Day Lily
蒟蒻麵疙瘩Konjac Noodle Pieces with Veggies
南瓜蒟蒻麵Konjac Noodles with Pumpkin Sauce
青椒炸醬麵Konjac Noodles with Green Pepper in Bean Sauce
茄汁蝴蝶麵Konjac Farelle in Tomato Sauce
味噌蒟蒻飯Konjac Rice with Miso
蒟蒻養生湯Konjac Body Supplement Soup

Konjac Shrimp Salad

Konjac Shrimp Salad

蒟蒻蝦沙拉

材料：（5~6人份）
蒟蒻蝦100克、低筋麵粉100克、水150c.c.、蘋果1顆、馬鈴薯泥480克、熟青豆仁2大匙、葡萄乾1大匙

醬料：
沙拉醬120克、橄欖油1大匙、鹽1小匙、黑胡椒粉適量

做法：
1.蒟蒻蝦加少許鹽浸泡30分鐘使其入味。
2.低筋麵粉加水調成漿，將蒟蒻蝦沾裹麵糊放入高溫油中炸1分鐘。
3.蘋果洗淨切丁，加少許醋水浸泡以避免氧化變黑。
4.將所有醬料加馬鈴薯泥調勻，依序放入蒟蒻蝦、青豆仁、葡萄乾及蘋果丁即可。

INGREDIENTS: 5-6 people
100g konjac shrimp,100g cake flour,150cc water, 1 apple, 480g mashed potato, 2T cooked peas, 1T raisins

SEASONINGS FOR MAKING DRESSING:
120g salad dressing, 1T olive oil, 1t salt, black pepper as needed

METHODS:
1. Marinate konjac shimp with a little salt for 30 minutes until the flavor is absorbed.
2. Combine cake flour with water to form batter, then dip konjac shrimp in batter thoroughly and deep-fry in smoking oil for 1 minute.
3. Rinse apple well and dice, then soak in a little vinegar water to prevent it from darkening due to oxidation.
4. Combine all the seasonings for the dressing well together, add mashed potato to mix. Then add konjac shrimp, peas, raisins and diced apple in order. Mix well and serve.

1.油炸蒟蒻必須高溫、快速，否則蒟蒻炸老了會影響口感，不想攝取過多熱量的人，可以省略掉油炸的步驟。
2.馬鈴薯和青豆仁都屬膳食纖維高的食物，吃了有飽足感而且又營養，一舉數得。
3.葡萄乾含有不少的膳食纖維及鐵質，可有效改善便祕及貧血的情形，是富營養價值的零食點心。

1. The oil temperature must be very high and the frying must be rapid when deep-frying konjac, or over-frying will affect its texture. If you want to reduce caloric intake, skip the deep-frying procedure..
2. Potatoes and peas are high fiber vegetables. Serving them fills the stomach and provides nutrients for the body. It repays handsomely the small investment of time and energy needed to prepare it.
3. Raisins are high in fiber and iron. High-nutrition snacks, they help with constipation and anemia.

Fresh Veggie Curry Konjac

Fresh Veggie 鮮蔬咖哩蒟蒻 Curry Konjac

材料：（5~6人份）
熟蒟蒻塊200克、青蘋果1顆、綠花椰菜100克、茄子1條、馬鈴薯1個、小蕃茄150克

調味料：
油1大匙、咖哩粉3大匙、椰漿240c.c.、牛奶240c.c.、鹽1/4小匙、糖1小匙

做法：
1.馬鈴薯去皮切成塊狀蒸熟。
2.小蕃茄、蘋果、茄子都去蒂洗乾淨切塊，綠花椰菜洗乾淨切成小朵備用。
3.1大匙熱油先將咖哩粉炒香，放入牛奶、椰漿及所有材料同煮至沸騰。
4.最後放入糖及鹽調味即可。

INGREDIENTS: 5-6 people
200g konjac pieces, 1 green apple, 100g broccoli, 1 egg-plant, 1 potato, 150g cherry tomatoes

SEASONINGS:
1T cooking oil, 3T curry powder, 240cc coconut milk, 240cc milk, 1/4t salt, 1t sugar

METHODS:
1.Peel potato and cut into pieces, then steam until done.
2.Remove stems from cherry tomatoes, apple and egg-plant, then rinse well and cut into pieces. Rinse broccoli and cut into small florets.
3.Heat 1T of cooking oil in wok and stir-fry curry powder until fragrant. Add milk, and coconut milk along with all the other ingredients. Bring mixture to boil.
4.Season with sugar and salt to taste. Serve.

1.許多小朋友會對蔬菜挑食，媽媽們不妨將蔬菜都放入咖哩醬汁中，做成美味的咖哩美食，可以使小朋友們吃下大量蔬菜，既營養又可獲得健康。
2.除了以上的新鮮蔬果外，像胡蘿蔔、青豆仁、洋蔥、秋葵等都可以加入咖哩醬汁中烹煮。

1.Many small kids are very picky about vegetables. I suggest mothers put the vegetables in curry sauce. Kids will love it, and get nutrition by eating lots of vegetables.
2.In addition to the fresh vegetables mentioned above, others like carrot, peas, onion or okra can be added to the curry sauce.

Konjac Scallop and Chestnuts

Konjac Scallop 栗燒蒟蒻干貝 and Chestnuts

材料：（5~6人份）
熟蒟蒻塊200克、熟栗子150克、青江菜8~10顆

調味料：
油1/2小匙、醬油3大匙、冰糖粉1大匙、水220c.c.

做法：
1.蒟蒻塊用圓模型壓成每個高度約1.5公分的干貝狀，將干貝浸泡在調味料中20~30分鐘讓干貝入味。
2.青江菜整顆洗淨，放入沸水燙熟後取出漂涼，瀝乾水分後排入盤中。
3.取一鍋子，倒入浸泡干貝的醬汁，先放入栗子，用小火慢燒7~8分鐘，再放入干貝一起燒至湯汁收乾熄火，最後倒入已排盤的青江菜中即可。

INGREDIENTS: 5-6 people
200g konjac pieces, 150g cooked chestnuts, 8-10 stalks baby bok choy

SEASONINGS:
1/2t cooking oil, 3T soy sauce, 1T rock sugar powder, 220cc water

METHODS:
1.Press each konjac piece with a round mold to make scallop shapes about 1.5 cm high, then soak them in well-mixed seasonings for about 20 to 30 minutes until the flavor is well absorbed.
2.Rinse the whole stalks of baby bok choy, then blanch in boiling water until done and drain well. Remove and arrange on the sides of the serving plate.
3.Cook the well-mixed seasonings from soaking konjac scallops in pan, add chestnuts and cook over low heat for about 7 to 8 minutes, then add konjac scallops. Continue cooking until the liquid is completely absorbed, remove from heat and pour in the center of the serving plate.

TIPS

在製作蒟蒻干貝時，可以用叉子在每個干貝上戳一戳，一方面加速去鹼，一方面加強入味的效果。

When preparing the konjac scallops, prick each one with a fork several times to release the alkalinity and to help it absorb the flavor.

Baked Stuffed Bell Peppers

Baked Stuffed 煎鑲甜椒 Bell Peppers

材料： （5~6人份）
蒟蒻丁100克、紅甜椒2顆、玉米粒100克、洋菇100克、嫩豆腐200克、起司粉1大匙

調味料：
油1/2小匙、鹽1/2小匙、黑胡椒粉1小匙、太白粉3大匙

做法：
1.甜椒洗淨橫切1/2，去籽備用。
2.將洋菇切碎粒，與蒟蒻丁、嫩豆腐、玉米粒加入調味料一起拌匀，放入油鍋中拌炒至8分熟，起鍋後加入太白粉拌匀。
3.每1/2個甜椒內塞入做法2.的餡料，放入烤箱中以180℃（356℉）烤8分鐘，取出後排盤。
4.在甜椒上撒起司粉即可食用。

INGREDIENTS: 5-6 people
100g konjac dices, 2 red bell peppers, 100g corn kernels, 100g button mushrooms, 200g tender tofu, 1T cheese powder

SEASONINGS:
1/2t cooking oil, 1/2t salt, 1t black pepper powder, 3T cornstarch

METHODS:
1. Rinse bell peppers well and cut horizontally in half, then discard the seeds.
2. To prepare filling: Chop button mushrooms finely, mix well with konjac dices, tender tofu, corn kernels along with seasonings. Stir-fry with oil in pan until 80% done, remove from heat and mix well with cornstarch.
3. Stuff the filling from method (2) inside each bell pepper half. Bake in oven at 180℃（356℉）for 8 minutes until done. Remove and arrange on serving plate.
4. Sprinkle the dish with cheese powder and serve.

1.用烤的食物少了許多「油」的熱量，很適合晚餐食用。
2.紅甜椒顏色漂亮，與任何食材搭配都能促進食慾，再加上甜椒含有生素C、β－胡蘿蔔素和蕃茄紅素，能防止身體的細胞受到自由基的傷害，延緩老化，稱得上是營養滿點的新時代健康食材。
3.吃洋菇會增加尿酸的含量，痛風患者勿食用。

1. Baked dish has no calories from oil and is perfect for supper.
2. Red bell pepper has very beautiful color. No matter what it is paired with, it can stimulate the appetite. Bell pepper contains vitamin C, carotene and lycopene. It prevents the body cells from the damage of the free radicals and slows down aging. It could be a called a 100% nutritious food.
3. Button mushroom increase uric acid, so people with gout should not eat it.

Flip-Over Pumpkin Konjac

Flip-Over
南瓜蒟蒻扣
Pumpkin Konjac

材料：（8~10人份）
白、灰蒟蒻各50克、南瓜1/2個（約200克）、傳統豆腐200克、枸杞10克、蒸熟蓮子30克、綠花椰菜1顆

調味料：
鹽1/2小匙、水120c.c.、高湯1大匙、太白粉3大匙

做法：
1.灰蒟蒻是加了少許黑芝麻粉做成的蒟蒻塊，做法參見p.59。
2.南瓜洗淨表皮，切半去籽，放入鍋中蒸約12~15分鐘，取出後用杓子挖出南瓜肉。
3.灰、白蒟蒻都切成薄片備用。
4.傳統豆腐用紗布擰去水分，與南瓜泥、鹽及2大匙太白粉拌勻。
5.取一中碗，抹上少許油，將蒟蒻片以灰、白相間的方式排入碗中，再將做法3.倒入，稍微壓緊後，以中火隔水蒸約10分鐘後取出扣入盤中。
6.綠花椰菜切成小朵，洗淨燙熟後排在南瓜泥盅周圍。
7.鍋中放入高湯、水、蓮子、枸杞，煮沸後用濕太白粉勾成薄芡淋在盤上即可。

INGREDIENTS: 8-10 people
50g each white and gray konjac, 1/2 pumpkin (about 200g), 200g traditional tofu, 10g lycium berries, 30g steamed lotus seeds, 1 head broccoli

SEASONINGS:
1/2t salt, 1T soup broth, 120cc water, 3T cornstarch

METHODS:
1. Gray konjac is konjac chunks with a little black sesame powder added, see methods on p.59.
2. Rinse the surface of pumpkin well, halve and discard seeds. Steam in steamer for about 12-15 minutes until done. Remove and scoop out the pumpkin flesh with a ladle or spoon.
3. Cut white and gray konjac into thin slices.
4. Squeeze out excess water from tofu with a cheese-cloth, then mix well with pumpkin, salt and 2T of corn-starch.
5. Grease a medium size bowl with oil, line white and gray konjac alternately at the bottom, then top with method (3). Press tight lightly and steam over medium heat in a double boiler for about 10 minutes until done. Remove and flip over on a serving plate.
6. Cut broccoli into small florets, rinse well and blanch until done. Arrange on the sides of the steamed pumpkin.
7. Heat soup broth, water, lotus seeds and lycium berries until boiling, thicken with cornstarch water and drizzle over the dish. Serve.

1.這道菜手工較複雜，需要花較多的時間。將菜倒扣起來很漂亮，偶有嘉賓來訪時，可以show一下手藝！
2.綠花椰菜含有豐富的抗癌成分，可多食用。
3.枸杞具有保護眼睛、消除疲勞、預防動脈硬化及防止老化等功效，好處多多。

1. This dish takes lots of skill and preparation time. Flipped over, the appearance is quite appealing. When you have guests over, use it to show off your skills.
2. Broccoli contains abundant anti-cancer nutrients, and should be served more often.
3. Among their numerous benefits Lycium berries protect the eyes, release fatigue, and prevent atherosclerosis and aging.

Konjac Shrimp over Burnt Rice

Konjac Shrimp over Burnt Rice

蝦仁鍋巴

材料：（5~6人份）
蒟蒻蝦150克、乾香菇30克、豆莢30克、胡蘿蔔30克、筍片30克、鍋巴100克

調味料：
油1小匙、醬油1/2小匙、鹽1/4小匙、香油1/4小匙、水120c.c.、太白粉水1大匙

做法：
1. 香菇泡軟切片；豆莢洗淨撕去纖維；胡蘿蔔切片。
2. 鍋中放油炒香菇，放入蒟蒻蝦、胡蘿蔔、豆莢等，加入水、鹽、醬油調味，再以中火煮開。
3. 加太白粉水勾成薄芡，撒入些許香油後熄火，趁熱將醬汁淋在鍋巴上即可。

INGREDIENTS: 5-6 people
150g konjac shrimp, 30g dried shiitake mushrooms, 30g pea pods, 30g carrot, 30g bamboo shoot slices, 100g burnt rice from the bottom of the cooker

SEASONINGS:
1t cooking oil, 1/2t soy sauce, 1/4t salt, 1/4t light sesame oil, 120cc water, 1T cornstarch water

METHODS:
1. Soak shiitake mushrooms in water until soft, then slice. Rinse pea pods and tear off strings from both sides. Cut carrot into thin slices.
2. Heat oil in wok and stir-fry shiitake mushrooms until fragrant. Add konjac shrimp, carrot slices, pea pods along with water, salt and soy sauce to taste. Bring to boil over medium heat.
3. Thicken with cornstarch water and drizzle with light sesame oil, then remove from heat. Drizzle over burnt rice while is still steaming. Serve.

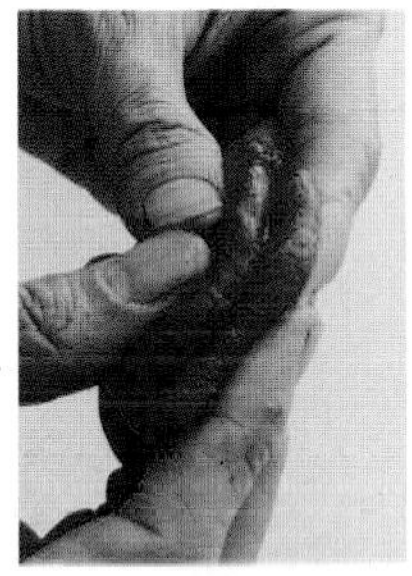

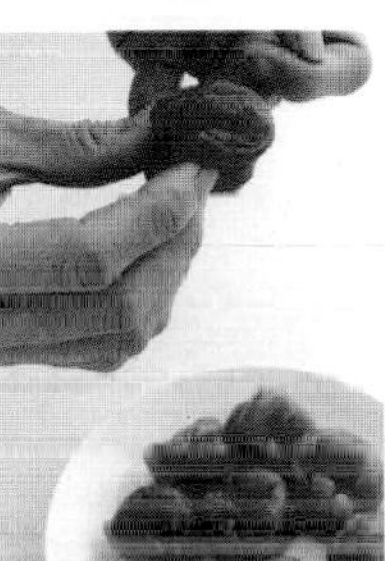

蒟蒻蝦做法參見P.14

1. 鍋巴在一般大型超市中都買得到，只要不受潮，不經油炸也能有酥脆的口感。
2. 這道菜完成後須儘快食用，否則鍋巴會被醬汁沾濕變得軟綿綿的。
3. 竹筍是含有豐富纖維質的食物，它可以預防及改善便祕，而且可以減少罹患大腸癌的機率。但竹筍纖維較粗，不可一次吃太多，以免妨礙消化。

1. Burnt rice can be purchased at large supermarkets. If preserved well, without contact with moisture, it will retain its crunchy texture without needing deep-frying.
2. This dish is better served right away, or the burnt rice will become soggy due to the stir-fried konjac shrimp mixture on top.
3. Bamboo shoots are high in fiber and prevent and improve constipation. It reduces cancer of the large intestine. However, the fiber is coarse and tough, do not eat too much at a time, or you might have trouble digesting.

Konjac Tofu Balls with Bok Choy

Konjac Tofu Balls with Bok Choy

蒟蒻燴福球

材料：（5~6人份）
熟蒟蒻塊200克、大白菜1/2顆、傳統豆腐1塊、荸薺8顆、辣椒絲少許、油1杯

調味料：
A.鹽1/4小匙、玉米粉2大匙、薑末少許、糖1小匙
B.醬油2小匙、糖1/4小匙

做法：
1.蒟蒻切小丁，荸薺拍碎備用。豆腐洗淨壓泥，用紗布擠出多餘水分，加入蒟蒻丁、荸薺和調味料A.做成豆腐球。
2.起油鍋燒至八分熱，放入豆腐球炸成金黃後取出。
3.倒出炸油，留約1大匙在鍋中，將洗淨切段的大白菜炒軟，放入豆腐球及調味料B.，轉小火燉煮10分鐘，盛盤撒上辣椒絲即可。

INGREDIENTS: 5-6 people
200g shredded konjac, 1/2 head bok choy, 1 piece traditional tofu, 8 waterchestnuts, shredded chili pepper as needed, 1C cooking oil

SEASONINGS:
A: 1/4t salt, 2T cornstarch/corn flour, minced ginger as needed, 1t sugar
B: 2t soy sauce, 1/4t sugar

METHODS:
1. Dice konjac well. Crush water chestnuts. Rinse tofu and mash, remove to cheesecloth and squeeze out the excess water. Mix well with water chestnuts and seasoning (A) in bowl. Divide into portions and make into tofu balls.
2. Heat oil in pan to 80% smoking, deep-fry tofu balls until golden and remove.
3. Remove the oil and retain 1T of oil in pan, stir-fry bok choy until soften, add tofu balls as well seasoning (B) and reduce heat to low. Cook for 10 minutes and remove to plate, then sprinkle with shredded chili pepper. Serve.

1.荸薺又叫「馬蹄」，含有維生素A、B1、B2、C和礦物質鈣、磷、鐵等，雖營養素多，但不可吃太多，否則容易腹漲。
2.荸薺咀嚼起來有種特殊的口感，可加在蝦鬆、水餃餡中，另有一番滋味！
3.大白菜含大量的水及粗纖維，可促進腸部蠕動、幫助消化，有利排便，預防便祕。
4.豆腐中含有高量的異黃酮素，它是一種抗氧化劑，多吃可減少罹患癌症的機會，還可降低血液中的膽固醇，預防老年痴呆症。

1. Water chestnuts contain vitamin A. B1. B2. C and minerals, such as calcium, phosphorus, and iron. Though they are very nutritious, do not eat too much, or you may have stomach gas pain.
2. Water chestnuts have a very special texture. Add to minced shrimp in lettuce cup or stuffed dumplings to give them a special flavor.
3. Bok choy is high in water and coarse fiber. It speeds bowel movement and helps digestion. It also smoothes intestinal movements and prevents constipation.
4. Tofu is high in isoflavones, a kind of antioxidant. Serving often can decrease the possibility of cancer, lowers blood cholesterol and prevents Alzheimer's disease.

Chinese Yam Konjac with Loofah and Day Lily

Chinese Yam Konjac with
山藥蒟蒻海棠煮
Loofah and Day Lily

材料：（8~10人份）
紫菜1張、山藥400克、絲瓜300克、蒟蒻蝦150克、金針10克、薑絲20克

調味料：
鹽1匙、油1小匙、濕太白粉1大匙、香油少許、水120c.c.

做法：
1.山藥去皮磨成泥，用1/2小匙鹽及1/2小匙油調味。
2.取一略有深度的碗，將紫菜鋪上去，再鋪山藥泥，放入電鍋中蒸約10分鐘。
3.金針去硬蒂、打結後泡軟。絲瓜去皮只取翠綠色的部分，中間白色的部分不用，切片備用。
4.起油鍋爆炒薑絲，放入絲瓜、金針、蒟蒻蝦略炒，加水燜熟、調味，以濕太白粉勾薄芡淋在蒸好的山藥泥上即可。

INGREDIENTS: 8-10 people
1 nori wrapper, 400g Chinese yam, 300g sponge loofah, 150g konjac shrimp, 10g dried yellow day lily flowers, 20g Shredded ginger

SEASONINGS:
1t salt, 1t oil, 1T cornstarch water, light sesame oil as needed, 120cc water

METHODS:
1.Peel Chinese yam and grind until well-mashed, then season with 1/2t of salt and 1/2t of oil to taste.
2.Line with a sheet of nori wrapper at the bottom of a deep bowl, then spread evenly with mashed Chinese yam. Remove to rice cooker and steam for about 10 minutes.
3.Remove stems from day lily flowers, tie into knots and soak in water until soft. Peel loofah, retaining only the green part, discard the white part, then slice.
4.Heat wok and stir-fry shredded ginger until fragrant. Add loofah, day lily flowers and konjac shrimp. Saute for a minute, then add water and cover. Simmer until done, season with seasonings and thicken with cornstarch water, then drizzle over the steamed mashed yam. Serve.

1.這道菜色澤漂亮、清淡爽口，大宴小酌都相當適合。
2.山藥是很受歡迎的養生食材，現在本土栽植的品種愈來愈好，而且幾乎一年四季都能嘗到，很幸福呢！
3.山藥含蛋白質、維生素C、糖、鈣、胺基酸、磷、鐵等重要物質，有益人體的健康，可多多食用。

1. The color of this dish is very pretty. The taste is light and delicious. It is quite suitable for entertaining friends at parties.
2. Chinese yam is a very popular nourishing food ingredient. We are lucky that the local version has improved greatly and is available year round!
3. Chinese yam contains many important minerals, such as protein, vitamin C, sugar, calcium, amino acid, phosphorus and iron. It is very beneficial to the health and should be eaten often.

Konjac Noodle Pieces with Veggies

Konjac Noodle Pieces
蒟蒻麵疙瘩
with Veggies

材料：（5~6人份）

a.蒟蒻粉40克、鹼粉4克、水1000c.c.

b.金針菇100克、香菇100克、胡蘿蔔50克、豆莢50克、水2000c.c.

調味料：

醬油3大匙、鹽1小匙、香菇粉1/2小匙、油1/2小匙

做法：

1.材料a.參照P.14做成蒟蒻塊，在室溫下靜置30~40分鐘。

2.煮一鍋水保持微微沸騰，用手抓蒟蒻成疙瘩狀投入水中煮。

3.待全部的蒟蒻都做成疙瘩後，一起再煮10分鐘，熄火倒去鹼水。

4.將蒟蒻麵疙瘩泡在乾淨的水中3~5小時，完全去鹼後才能開始烹調。

5.起油鍋炒香配料b.，加入水調味，放入麵疙瘩續煮至沸騰即可。

INGREDIENTS: 5-6 people

a. 40g konjac powder, 4g alakine powder, 1000cc water

b. 100g velvet foot mushrooms, 100g shiitake mushrooms, 50g carrot, 50g pea pods, 2000cc water

SEASONINGS:

3T soy sauce, 1t salt, 1/2t shiitake mushroom powder, 1/2t cooking oil

METHODS:

1.Combine ingredient (a) together and follow the steps at the Basic Methods (P.14) for Preparing konjac. Let sit at room temperature for 30 to 40 minutes.

2.Bring a pot of water to a boil and maintaining the water temperature. Meanwhile, tear konjac into pieces and throw into the water to cook.

3.Finish all the konjac pieces and continue cooking for 10 minutes longer, then remove from heat and discard the water.

4.Soak konjac pieces in clean water for 3 to 5 hours to get rid of the alkali flavor completely before starting to cook.

5.Heat oil in wok and stir-fry all the ingredients (b) until fragrant. Add a little water to mix, then add konjac pieces to mix. Bring to a boil and remove. Serve.

1. 凝結成塊的蒟蒻在尚未加熱前，就像黏土一樣，可以做成任何喜歡的形狀，做成麵疙瘩是最簡單的一種，也可以做成炒麵疙瘩、紅燒麵疙瘩等……真的很有趣，對嗎？
2. 對於想減肥的人來說，可以利用這道蒟蒻麵疙瘩代替主食，不僅熱量低、吃後有飽足感，加上多種新鮮蔬菜的營養，輕鬆兼顧瘦身及營養均衡。

1. Before setting, uncooked konjac feels like play dough, and can be shaped into anything desired. In this recipe, you can mold into other shapes, but tearing it into pieces is the easiest one. Konjac pieces can also be prepared as formal dishes, like stir-fried konjac pieces, or red-cooked konjac pieces. Isn't that fun?
2. For those who are on a diet, use konjac pieces in the main dish. It is not only low in calories, but fills the stomach. Coupled with the variety of nutrition from the vegetables, you can keep fit and obtained balanced nutrition easily.

Konjac Noodles with Pumpkin Sauce

Konjac Noodles with Pumpkin Sauce

南瓜蒟蒻麵

材料：（3~4人份）
蒟蒻麵450克、蘑菇180克、松茸菇150克、南瓜500克、馬鈴薯100克、巴西利末少許

調味料：
奶油1大匙、鮮奶500克、水100克、鹽1/4小匙、黑胡椒適量

做法：
1.待生蒟蒻塊不沾手時，取一小塊用擀麵棍成片狀再切長條，然後放入水中煮沸後取出，泡水3~5小時去鹼，即成蒟蒻麵。
2.南瓜去籽不去皮、馬鈴薯去皮，切成小塊蒸熟。
3.將蒸熟的馬鈴薯塊及1/2的南瓜塊和鮮奶、水打成稠狀。
4.蘑菇切碎用奶油炒香，將松茸菇、南瓜汁一起倒進鍋中煮，煮至約90°C（194°F）時放入蒟蒻麵、另1/2南瓜塊，以鹽、黑胡椒調味再煮至沸騰，撒上些許巴西利末即可。

INGREDIENTS: 3-4 people
450g konjac noodles, 180g mushrooms, 150g pine mushrooms, 500g pumpkins, 100g potatoes, parsley flakes as needed

SEASONINGS:
1T butter, 500g milk, 100g water, 1/4t salt, black pepper as needed

METHODS:
1. To prepare konjac noodle - Wait until the konjac chunks are not sticky, remove a little and roll flat with a rolling pin, then cut into long strips. Cook in boiling water until done and remove to soak in water for 3 to 5 hours to remove the alkali.
2. Retain the skin and discard the seeds from pumpkin. Peel potatoes and cut into small pieces, then steam until done in the steamer.
3. Combine the steamed potatoes with 1/2 of the pumpkin along with milk and water in blender. Blend until thickened.
4. Chop mushrooms until finely and stir-fry with butter until fragrant. Transfer to pan along with pine mushroom and pumpkin liquid added. Cook until the temperature reaches about 90°C（194°F）, add konjac noodles, and another 1/2 of the pumpkin pieces along with salt and black pepper to taste. Continue heating until boiling. Remove and serve.

1.金黃色的南瓜就像陽光一樣令人愉悅，拿來入菜不僅營養，更增添料理的顏色！
2.蒟蒻麵晚點兒放是為了不讓它口感過硬，如果對牛奶過敏，可以用原味豆漿代替。

1. Golden pumpkin is just like the sunny sun, filled with joy. Using it in dishes is not only nutritious, but also enhances the color of the dish.
2. Konjac noodles are added at the end to prevent the texture from being too tough. If you are allergic to milk, use plain soy milk instead.

Konjac Noodles with Green Pepper in Bean Sauce

Konjac Noodles with
青椒炸醬炒麵
Green Pepper in Bean Sauce

材料：（3~4人份）
蒟蒻麵300克、蒟蒻丁200克、豆乾丁50克、毛豆50克、乾香菇20克、青椒150克

調味料：
豆瓣醬4大匙、甜麵醬1大匙、糖1小匙、白胡椒粉1/4小匙、水100c.c.、油2大匙

做法：
1.乾香菇泡軟，與豆乾丁、毛豆一起切碎。
2.鍋中放油，將香菇丁、毛豆丁等炒香，放入其他調味料以小火煮7~8分鐘。
3.放入蒟蒻丁、青椒絲、蒟蒻麵一起拌炒5分鐘即可。

INGREDIENTS: 3-4 people
300g konjac noodles, 200g konjac cubes, 50g dried diced pressed tofu, 50g young soybeans, 20g dried shiitake mushrooms, 150g green pepper

SEASONINGS:
4T soybean paste, 1T sweet bean paste, 1t sugar, 1/4t white pepper, 100cc water, 2T cooking oil

METHODS:
1. Soak dried shiitake mushrooms in water until soft, then dice together with pressed tofu and young soybeans.
2. Heat oil in wok, stir-fry diced shiitake mushroom and young soybeans until fragrant. Add seasonings to taste. Continue cooking over low heat for about 7 to 8 minutes.
3. Add diced konjac, shredded green pepper and konjac noodles to mix. Saute for about 5 minutes and remove to serve.

TIPS

1.加了蒟蒻丁的炸醬，也可以用來拌飯、做成乾麵等。蒟蒻麵的最大作用就是增加飽足感，搭配大量的蔬菜就能飽餐一頓了！
2.對喜歡吃麵食又怕發胖的人來說，以蒟蒻麵取代一般麵條，不僅熱量攝取低，美味也不打折扣。
3.青椒可以生吃也可以烹調後再吃，但生吃較能攝取到更多營養素。

1. This bean sauce mixture with diced konjac added can be mixed with rice or noodles. The function of Konjac noodles is to increase the fullness feeling. With a large quantity of vegetables added, your stomach will be filled with satisfying food.
2. For those who enjoy noodle dishes, yet are afraid of high calories, using konjac noodles instead of ordinary noodles not only lowers the calories, but still offers delicious flavor.
3. Green pepper can be served either raw or cooked, but raw is more nutritious.

Konjac Farelle in Tomato Sauce

Konjac Farelle 茄汁蝴蝶麵 in Tomato Sauce

材料：（5~6人份）
蒟蒻蝴蝶麵600克、新鮮紅蕃茄400克、蕃茄汁250c.c.、起司粉2大匙、俄力岡1/2小匙、鹽1/2小匙、黑胡椒粉1/4小匙、油1/2小匙

做法：
1.蕃茄去蒂洗淨，切成小丁。
2.起油鍋將紅蕃茄炒軟，加入蕃茄汁、蝴蝶麵、鹽、俄力岡、黑胡椒粉等調味料，持續以中火加熱煮至沸騰。
3.盛盤後撒上起司粉即可。

INGREDIENTS: 5-6 people
600g konjac farelle, 400g fresh red tomatoes, 250cc tomato juice, 2T cheese powder, 1/2t oregano, 1/2t salt, 1/4t black pepper, 1/2t cooking oil

METHODS:
1. Remove stems from tomatoes and rinse well, then cut into small dices.
2. Heat oil in wok, stlr-fry tomato dices until soft, add tomato juice and konjac along with salt, oregano and black pepper to taste. Continue heating over medium until boiling.
3. Remove to serving plate and sprinkle with cheese powder. Serve.

蒟蒻蝴蝶麵做法
參見P.17

1.簡單的材料加上風味特殊的天然香料，就能變化出異國風味。
2.蕃茄的種類很多，綠一些的比較有蕃茄味，紅一些的則顏色比較好看，建議不妨兩種一起用。

1. Simple ingredients and natural herbs with a special flavor can create a foreign flavor.
2. There are many kinds of tomatoes. Green ones have a heavier tomato flavor, while red ones are beautiful in color. I suggest using them both.

Konjac Rice with Miso

Konjac Rice 味噌蒟蒻飯 with Miso

材料：（3~4人份）
蒟蒻米200克、生米180克、乾香菇4朵、芋頭100克、芹菜末1大匙

調味料：
油1小匙、淡色味噌50克、鹽1/4小匙、黑胡椒1/2小匙、無糖豆漿240c.c.

做法：
1.蒟蒻塊加水用果汁機打碎，即成蒟蒻米。
2.香菇泡軟去蒂每朵切4片，芋頭去皮切小塊。
3.用油將香菇炒香，加入生米、蒟蒻米再炒2分鐘。
4.放入芋頭、豆漿、味噌、鹽、黑胡椒等均勻拌炒2分鐘。
5.轉小火燜煮10分鐘後攪拌一下，續煮至湯汁收乾。
6.將飯裝盤，撒上些芹菜末即可。

INGREDIENTS: 3-4 people
200g konjac rice, 180g uncooked rice, 4 dried shiitake mushrooms, 100g taro, 1T diced celery

SEASONINGS:
1t cooking oil, 50g light miso, 1/4t salt, 1/2t black pepper, 240cc unsweetened soybean milk

METHODS:
1. To prepare konjac rice - Blend konjac chunks with water in blender until mashed.
2. Soak shiitake mushrooms in water until soft, remove and cut each into 4 equal slices. Peel taro and cut into small pieces.
3. Heat oil in wok and stir-fry shiitake mushroom slices until fragrant. Add uncooked rice and konjac rice. Continue stirring for 2 minutes longer.
4. Add taro and soybean milk along with miso, salt and black pepper to taste. Saute rapidly for 2 minutes until evenly mixed.
5. Reduce heat to low and simmer for 10 minutes, then stir to mix. Continue cooking until the liquid is well absorbed.
6. Transfer to serving plate and sprinkle with diced celery. Serve.

1.蒟蒻塊加水用果汁機打碎就成了蒟蒻米。蒟蒻米飯能夠增加飽足感，用蒟蒻加一般米煮飯時，只要計算一般米所需的水分就可以了。
2.味噌的用途很廣，除了最常見的味噌湯外，還可以用來醃食材或烤魚時塗抹的沾料，或純粹用做沾醬來沾東西吃。

1. Combine konjac with water in blender and blend until fine to make konjac rice. Konjac rice increases the feeling of fullness. When konjac rice is added to rice, just measure the water by counting the cups of the uncooked rice.
2. Miso has a large variety of usages. In addition to the common miso soup, it can be used in marinades or sauces for spreading on the roasted fish, or simply used in dipping sauces.

Konjac Body Supplement Soup

Konjac Body 蒟蒻養生湯 Supplement Soup

材料：（5~6人份）
芝麻蒟蒻塊200克、山藥200克、猴頭菇30克、生腰果100克、枸杞20克、當歸1片、黃耆20克、紅棗15顆

調味料：
鹽2~3大匙、水2000c.c.

做法：
1. 猴頭菇泡水3小時（每小時換1次水），撕成小塊再用鹽水煮開，擠乾水分備用。
2. 山藥去皮切成小塊，蒟蒻切成適口大小，備用。
3. 取一深鍋，放入水、當歸、黃耆、紅棗、生腰果、猴頭菇等，煮沸後轉小火續煮30分鐘再轉中火。
4. 山藥、蒟蒻、枸杞一起放入鍋中再煮10分鐘，最後加鹽調味即可。

INGREDIENTS: 5-6 people
200g konjac sesame pieces, 200g Chinese yam, 30g monkey head mushroom, 100g raw cashews, 20g lycium berries, 1 slice Chinese angelica root, 20g astragalus root, 15 red jujubes

SEASONINGS:
2-3T salt, 2000cc water

METHODS:
1. Soak monkey head mushrooms in water for 3 hours (changing water every hour), then tear into small pieces and bring to boil in salt water. Remove and squeeze out the excess water.
2. Peel Chinese yam and cut into small pieces. Cut konjac into serving sizes.
3. Prepare a stewing pot, add water, Chinese angelica root, astragalus root, red jujubes, raw cashews and monkey head mushrooms. Bring to a boil first, then reduce heat to low and continue cooking for 30 minutes longer, then increase heat to medium.
4. Add Chinese yam, konjac and lycium berries, continue cooking for another 10 minutes. Season with salt to taste. Remove and serve.

1. 這道湯品味美甘甜，滋味無窮。生腰果煮成湯的口感一定會讓你驚訝；蒟蒻最後才放，是為防止它煮太久而變硬。當然，事先浸泡蒟蒻入味的手續是不能省略的喲！
2. 黃耆有抗老化、防癌及免疫作用；枸杞可明目，可泡枸杞茶、做成枸杞麵包或枸杞粥來吃；猴頭菇則可助消化、抗癌，通常在南北貨市場都可以買得到。
3. 紅棗可補血，可泡枸杞紅棗茶來喝；當歸對消化系統及呼吸系統有所助益；山藥有圓形、長形及掌形等種類，以長形山藥較受歡迎，利用也較多，但各種類的山藥都有養生功效，具高營養價值。

1. This soup dish is sweet and delicious with unlimited flavor. You will be surprised to try the texture of the raw cashews in soup. Add konjac at the end to prevent it from toughening after cooking too long. Of course, do not forget to marinate the konjac so the flavor can be absorbed.
2. Astragalus root fights aging, prevents cancer, and builds immunity. Lycium berries clarify the eyes, and can be made into tea, bread or porridge. Monkey head mushrooms help digestion, fight cancer and can be found in any dried good markets.
3. Red jujubes are good for the blood, and can be made into tea with lycium berries added. Chinese angelica root benefits the digestion and respiratory system. The are many varieties of Chinese yam, including round, long and palm shapes. The long shape is the most popular and commonly-used. However, each type of yam has a nourishing effect and all are valuable for their nutrition.

和你快樂品味生活

北市基隆路二段13-1號3樓　http://redbook.com.tw

COOK50系列　基礎廚藝教室

編號	書名	作者	定價
COOK50001	做西點最簡單	賴淑萍著	定價280元
COOK50002	西點麵包烘焙教室－乙丙級烘焙食品技術士考照專書	陳鴻霆、吳美珠著	定價480元
COOK50003	酒神的廚房	劉令儀著	定價280元
COOK50004	酒香入廚房	劉令儀著	定價280元
COOK50005	烤箱點心百分百	梁淑嫈著	定價320元
COOK50006	烤箱料理百分百	梁淑嫈著	定價280元
COOK50007	愛戀香料菜	李櫻瑛著	定價280元
COOK50008	好做又好吃的低卡點心	金一鳴著	定價280元
COOK50009	今天吃什麼－家常美食100道	梁淑嫈著	定價280元
COOK50010	好做又好吃的手工麵包－最受歡迎麵包大集合	陳智達著	定價320元
COOK50011	做西點最快樂	賴淑萍著	定價300元
COOK50012	心凍小品百分百－果凍・布丁（中英對照）	梁淑嫈著	定價280元
COOK50013	我愛沙拉－50種沙拉・50種醬汁（中英對照）	金一鳴著	定價280元
COOK50014	看書就會做點心－第一次做西點就OK	林舜華著	定價280元
COOK50015	花枝家族－透抽軟翅魷魚花枝 章魚小卷大集合	邱筑婷著	定價280元
COOK50016	做菜給老公吃－小倆口簡便省錢健康浪漫餐99道	劉令儀著	定價280元
COOK50017	下飯ㄟ菜－讓你胃口大開的60道料理	邱筑婷著	定價280元
COOK50018	烤箱宴客菜－輕鬆漂亮做佳餚（中英對照）	梁淑嫈著	定價280元
COOK50019	3分鐘減脂美容茶－65種調理養生良方	楊錦華著	定價280元
COOK50020	中菜烹飪教室－乙丙級中餐技術士考照專書	張政智著	定價480元
COOK50021	芋仔蕃薯－超好吃的芋頭地瓜點心料理	梁淑嫈著	定價280元
COOK50022	每日1,000Kcal瘦身餐－88道健康窈窕料理	黃苡菱著	定價280元
COOK50023	一根雞腿－玩出53種雞腿料理	林美慧著	定價280元
COOK50024	3分鐘美白塑身茶－65種優質調養良方	楊錦華著	定價280元
COOK50025	下酒ㄟ菜－60道好口味小菜	蔡萬利著	定價280元
COOK50026	一碗麵－湯麵乾麵異國麵60道	趙柏淯著	定價280元
COOK50027	不失敗西點教室－最容易成功的50道配方	安 妮著	定價320元
COOK50028	絞肉の料理－玩出55道絞肉好風味	林美慧著	定價280元
COOK50029	電鍋菜最簡單－50道好吃又養生的電鍋佳餚	梁淑嫈著	定價280元
COOK50030	麵包店點心自己做－最受歡迎的50道點心	游純雄著	定價280元
COOK50031	一碗飯－炒飯健康飯異國飯60道	趙柏淯著	定價280元
COOK50032	纖瘦蔬菜湯－美麗健康、免疫防癌蔬菜湯	趙思姿著	定價280元
COOK50033	小朋友最愛吃的菜－88道好做又好吃的料理點心	林美慧著	定價280元
COOK50034	新手烘焙最簡單－超詳細的材料器具全介紹	吳美珠著	定價350元
COOK50035	自然吃・健康補－60道省錢全家補菜單	林美慧著	定價280元
COOK50036	有機飲食的第一本書－70道新世紀保健食譜	陳秋香著	定價280元

TEL：(02) 2345-3868　　FAX：(02) 2345-3828

COOK50037	**靚補**－60道美白瘦身、調經豐胸食譜	李家雄、郭月英著	定價280元
COOK50038	**寶寶最愛吃的營養副食品**－4個月～2歲嬰幼兒食譜	王安琪著	定價280元
COOK50039	**來塊餅**－發麵燙麵異國點心70道	趙柏淯著	定價300元
COOK50040	**義大利麵食精華**－從專業到家常的全方位密笈	黎俞君著	定價300元
COOK50041	**小朋友最愛吃的冰品飲料**	梁淑嫈著	定價260元
COOK50042	**開店寶典**－147道創業必學經典飲料	蔣馥安著	定價350元
COOK50043	**釀一瓶自己的酒**－氣泡酒、水果酒、乾果酒	錢薇著	定價320元
COOK50044	**燉補大全**－超人氣・最經典，吃補不求人	李阿樹著	定價280元
COOK50045	**餅乾・巧克力**－超簡單・最好做	吳美珠著	定價280元
COOK50046	**一條魚**－1魚3吃72變	林美慧著	定價280元

TASTER系列 吃吃看流行飲品

TASTER001	**冰砂大全**－112道最流行的冰砂	蔣馥安著	特價199元
TASTER002	**百變紅茶**－112道最受歡迎的紅茶・奶茶	蔣馥安著	定價230元
TASTER003	**清瘦蔬果汁**－112道變瘦變漂亮的果汁	蔣馥安著	特價169元
TASTER004	**咖啡經典**－113道不可錯過的冰熱咖啡	蔣馥安著	定價280元
TASTER005	**瘦身美人茶**－90道超強效減脂茶	洪依蘭著	定價199元
TASTER006	**養生下午茶**－70道美容瘦身和調養的飲料和點心	洪偉峻著	定價230元
TASTER007	**花茶物語**－109道單方複方調味花草茶	金一鳴著	定價230元
TASTER008	**上班族精力茶**－減壓調養、增加活力的嚴選好茶	楊錦華著	特價199元
TASTER009	**纖瘦醋**－瘦身健康醋DIY	徐　因著	特價199元

Quick系列

QUICK001	**5分鐘低卡小菜**－簡單、夠味、經點小菜113道	林美慧著	特價199元
QUICK002	**10分鐘家常快炒**－簡單、經濟、方便菜100道	林美慧著	特價199元
QUICK003	**美人粥**－纖瘦、美顏、優質粥品65道	林美慧著	定價230元
QUICK004	**美人的蕃茄廚房**－料理・點心・果汁・面膜DIY	王安琪著	特價169元
QUICK005	**懶人麵**－涼麵、乾拌麵、湯麵、流行麵70道	林美慧著	特價199元
QUICK006	**CHEESE！起司蛋糕**－輕鬆做乳酪點心和抹醬	日出大地著	定價230元
QUICK007	**懶人鍋**－快手鍋、流行鍋、家常鍋、養生鍋70道	林美慧著	特價199元
QUICK008	**隨手做義大利麵・焗烤**－最簡單、變化多的義式料理70道	洪嘉妤著	定價199元

輕鬆做系列 簡單最好做

輕鬆做001	**涼涼的點心**	喬媽媽著	特價99元
輕鬆做002	**健康優格DIY**	陳小燕、楊三連著	定價150元

國家圖書館出版品預行編目資料

蒟蒻纖瘦健康吃 高纖 ・低卡・最好做
／齊美玲著.--初版.--
台北市：朱雀文化，2004〔民93〕
面；公分.--（Cook50；47）
ISBN 986-7544-13-7（平裝）
1.食譜
427.1　　　93008283

COOK50047

蒟蒻纖瘦健康吃

高纖・低卡・最好做

作　者　齊美玲
翻　譯　施如瑛
攝　影　徐博宇
美術設計　鄭雅惠
編　輯　郝喜每
發 行 人　莫少閒
出 版 者　朱雀文化事業有限公司
地　址　北市基隆路二段13-1號3樓
電　話　02-2345-3868
傳　真　02-2345-3828
劃撥帳號　19234566 朱雀文化事業有限公司
e-mail　redbook@ms26.hinet.net
網　址　http://redbook.com.tw
總 經 銷　展智文化事業股份有限公司
I S B N　986-7544-13-7
初版一刷　2004.06

定　價　280元
出版登記　北市業字第1403號